Marine Biology

Microalgae and Man

MARINE BIOLOGY

Additional books in this series can be found on Nova's website under the Series tab.

Additional e-books in this series can be found on Nova's website under the e-book tab.

MARINE BIOLOGY

MICROALGAE AND MAN

DILWYN J. GRIFFITHS

NOVA BIOMEDICAL

New York

For permission to use material from this book please contact us:
Telephone 631-231-7269; Fax 631-231-8175
Web Site: http://www.novapublishers.com

Library of Congress Cataloging-in-Publication Data

Griffiths, Dilwyn J.
Microalgae and man / Dilwyn J. Griffiths (School of Biological Sciences, James Cook University, Queensland, Australia).
pages cm
Includes bibliographical references and index.
ISBN 978-1-62808-215-9 (hardcover)
1. Microalgae--Industrial applications. I. Title.
TP248.27.A46G75 2013
579.8--dc23

2013020304

Published by Nova Science Publishers, Inc. † New York

Contents

Preface

The microalgae, with representatives in all but one of the major algal divisions, are an important component of the biota of the world's aquatic environments. They include the phytoplankton that are responsible for a large proportion of the primary production of all marine and freshwater bodies. They occur at the base of the food chains upon which the world's fisheries industries depend, and are a vital component in the numerous aquaculture projects upon which the world will increasingly come to rely for a large portion of its protein requirement. Their use for the mass production of stock feed and for direct human consumption, already being practised in many parts of the world, is likely to become of increasing importance in the future as is the exploitation of their ability to serve as a source of key metabolites in the food industry and in the synthesis of a range of other high-value products. The mass culture of microalgae under controlled conditions is also under consideration for its potential to provide an alternative source of biomass and for the production of biofuels, such as biodiesel, that does not compete for land that can be more profitably used for the production of traditional food crops. The potential of microalgae as a commercial source of another energy source, hydrogen, is also under investigation with a view to exploiting the ability of some microalgae, unique among oxygenic photosynthetic organisms, to release hydrogen gas produced from the photolytic splitting of water. The proven role of microalgae in wastewater treatment and in various environmental remediation processes as well as their potential contribution as a key component of a range of carbon-capture schemes is also likely to become of increasing importance in the future.

This book surveys our current understanding of those aspects of the biology of microalgae which constitute the basis of the range of practical applications now under consideration for their potential contribution to human health and well-being. The focus is largely on physiological and biochemical processes of microalgae as they are currently known, with the aim of providing some of the basic background information against which present and proposed future developments can be viewed. Many of the latter developments, if they are to be successful, will require collaboration of engineers, process biochemists and microbiologists as well as those capable of assessing economic and environmental considerations. It is hoped that this book, will provide for such workers, for students and for the lay-person, an overview of some of the relevant basic biology of the microalgae, highlighting their metabolic flexibility and their vast potential as a valuable resource that is yet to be fully realised.

Acknowledgments

I thank the following for permission to include figures redrawn and modified from previously published work:

Elsevier (Figure 2, Figure 10),
Informa Healthcare (Figure 12),
Inter-Research (Figure 8, Figure 15),
Kluwer Academic Publishers (Figure 13),
Nature Publishing Group (Figure 3),
Oxford University Press (Figure 1, Figure 4),
Springer (Figure 5–first published by Kluwer Academic Publishers, Figure 6–first published by Pleiades Publishing Ltd., Figure 7, Figure 11, Figure 14–first published by Springer Science + Business Media, Figure 16-first published by Springer/Dordrecht).
Taylor & Francis (Figure 1).

I also thank Bryn Hughes and Owen Hughes for their valuable assistance in preparing the figures for this work.

Finally, I thank my wife, Elen, for her unfailing support and encouragement. This book is dedicated to her.

Introduction

The microalgae constitute an arbitrary grouping of the smaller microscopic and mostly unicellular photosynthetic organisms, many of which have much in common with some other non-photosynthesising unicellular organisms, such as the slime moulds, various ciliates and other "protozoa". According to some classification schemes they are placed with these other groups in the kingdom Protista (Corliss, 1984) thus separating the microalgae from the other larger (multicellular) algae. The microalgae are, in fact, represented in all but one of the commonly recognised phyla (divisions) of algae, namely, the chlorophyta (green algae), chrysophyta, cryptophyta, dinophyta (dinoflagellates), euglenophyta (euglenoids), glaucophyta, haptophyta (includes the coccolithophorids which in some schemes are placed with the chrysophyta), ochrophyta (diatoms or bacillariophyta) and rhodophyta (red algae). Another group traditionally included with the microalgae is the cyanobacteria although, like the photosynthetic bacteria, they are prokaryotic and taxonomically belong in a different kingdom, the Monera. Including the cyanobacteria with the true microalgae recognises some basic similarities of ecology and physiology, in particular their common possession of an oxygenic photosynthetic mechanism. The microalgae are also generally taken to include colonial forms in which otherwise independent individuals, when they divide, produce daughter cells that tend to remain together in a loose association often for a number of further divisions to yield coenobia. Some of the simpler filamentous forms are also included with the microalgae.

The microalgae, therefore, constitute a polyphyletic group of some 25,000 morphologically defined species. They are often the most dominant organisms in aquatic ecosystems, both marine and freshwater, where, as phytoplankton, they occupy an important position as primary producers at the base of the food chains. They have long been recognised for their contributions to a range of human activities and for their impact upon various aspects of human health and wellbeing. The role of microalgae in sustaining natural fisheries world wide has, in recent decades been considerably expanded through their use in aquaculture where they often provide a more economic, and for certain cultured fish, crustacean and shell-fish species at certain critical stages of their development, an indispensable food source. The relative ease with which certain of the microalgae can be grown in culture under controlled conditions has led to their being extensively used for biomass production, so far largely at the pilot-plant level but increasingly on a large scale in ponds, raceways and various forms of photobioreactor systems. They are increasingly being used to supplement the more traditional forms of stock feed and even as a source of essential components of human diets.

Current concerns over the environmental impacts and the likely increasing costs of our ever-growing demands for fossil fuels to meet our energy requirements has focussed attention on microalgae as possible sources of replacement fuels. Compared with other biofuel sources currently in use, such as corn, soybean or sugar cane, microalgae have the advantage of being amenable to culture on non-arable land and in fresh and salt- water environments. Moreover, they are capable of rapid growth and the entire cellular mass can, under suitable conditions and with appropriate technology, be harvested virtually continuously. The ability of many microalgal species to supplement photosynthetic production by using a range of organic substrates as alternative sources of carbon and energy offers the possibility of linking their biomass production with effluent and waste-water treatment so offsetting some of the costs of the necessary plant and equipment. The ability of microalgae to grow in media enriched with CO_2 (up to 10%) offers the possibility of enhanced biomass production while, at the same time, providing a possible mechanism for CO_2 capture from flue-gas emissions from a range of industrial sources. Hydrogen production, well known for many years as a unique feature of microalgal photosynthesis, is now receiving considerable attention in the search for alternative sources of energy. The practicality of harnessing this feature of microalgal biochemistry on a large scale, as is the case with many other proposed applications, requires careful consideration and costing. It is hoped that this book, through its survey of the contribution of microalgae to human health and well-being, and through its coverage of the relevant aspects of microalgal biology underpinning these practical applications, will provide a background against which current and future developments in this exciting field of biology/biotechnology can be appreciated.

Chapter I

Microalgae and Global Biogeochemistry

Microalgae, through their dominance of the primary producers of the world's oceans and lakes, have throughout geological history been a major influence on global biogeochemical cycles. The nature and extent of this influence is well illustrated by the vast chalk deposits that are widespread across most of northern Europe (with equivalents around the world) which are essentially the preserved calcareous remains of coccolithophorids (Haptophyta). Other indicators of microalgal biogeological impacts are the silica deposits (diatomaceous earths), the products of the silica metabolism of numerous diatom species and the fossilised cellulose remains of dinoflagellates. The cell walls of dead diatoms can, under suitable conditions, accumulate on the sea floor to leave immense deposits of silica, in some places over 1,000m thick as in the Seymour Islands in the eastern Antarctic Peninsula (Sims et al., 2006). The contribution of microalgae to the world's deposits of various forms of fossil fuels, already well recognised, is likely to become of increasing significance as alternative source deposits such as the vast reserves of oil shales that are now being considered for commercial exploitation. Lamosites, derived largely from lacustrine planktonic micralgae, occur as vast oil-shale deposits at Green River in western United States and in a number of Tertiary lacustrine deposits in Queensland, Australia. The former is estimated to contain 215 billion tons of in-place shale oil (1.5 trillion U.S. barrels) (Dyni, 2003).

The contribution of various categories of microalgae to global biogeochemistry over different geological periods is evident from the fossil record as reported by Brasier (1980) based on the original data of Tappan & Loeblich (1973). Such records are necessarily a very selective and inevitably incomplete representation of the microalgal community of a particular period and a particular locality since it comprises only those species that have cells or cellular fragments resistant to the various processes of degradation and compression that occur during sedimentation. The record left by calcareous scales of coccolithophorids and the siliceous cell walls of the diatoms places the origin of modern marine eukaryotic phytoplankton in the Triassic (Lipps, 1970) (>200 x 10^6 years/Ma BP). Much of today's plankton form three major protistan lineages: the heterokonts, the dinoflagellates and the haptophytes. The first named (recognised by the possession of two heterodynamic and characteristically structured flagellae) include a number of pigmented divisions such as diatoms, pelagophytes, silicoflagellates, chrysophytes, synurophytes and some multicellular algal groups (van den Hoek et al. 1995).

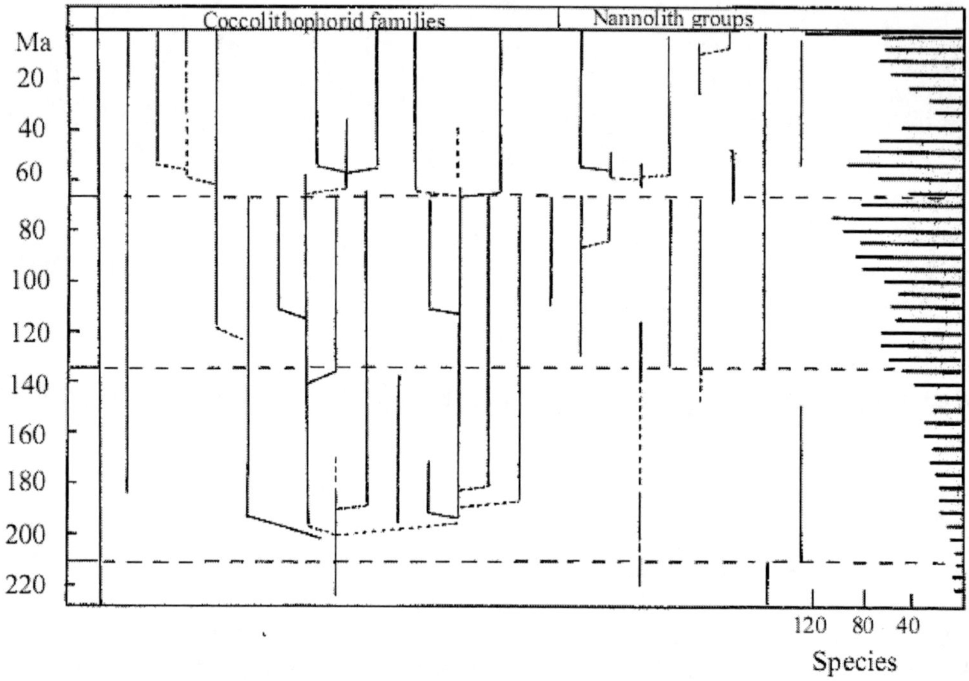

Figure 1. Presence (indicated by vertical lines with presumed evolutionary relationships) of identifiable (but not here named) coccolithophorid families and nannolith groups in relation to biostratigraphic age (0-220 x 10^6 years/Ma BP). The horizontal broken lines indicate the global Cretaceous/Tertiary, the Jurassic/Cretaceous and the Triassic/Permian extinction boundaries. Horizontal lines on the right-hand panel indicate total number of species in 5Ma intervals. (Redrawn, with permission, from Bown et al. (1991) as emended in Figure 20.3 by Young et al. (1994)).

The dinoflagellates and the haptophyta contain both autotrophic and heterotrophic members and many genera capable of adapting mixotrophic/phagotrophic forms of nutrition (Jones et al. 1994). Microalgae, through the presumed participation of their ancestral forms in the evolutionary phagotrophic event(s) leading to plastid formation (Bhattacharya & Medlin, 1995) and the consequent effects on photosynthetic production, have clearly played a fundamental role in shaping the global bioshere. The central importance of microalgae in global events over geological time is suggested by the strong correlation between the microalgal fossil record and the well-documented extinction events. It is likely that the two major post-Palaeozoic mass extinction events could have been largely driven, at least in oceanic regions, by the eradication of key microalgal groups, perhaps through effects on primary production or on calcification (Figure 1). Evidence that the divergence and subsequent radiation of the heterokontic microalgae may have been forced by the Permian-Triassic extinction event has been reviewed by Medlin (2011).

Coccolithophorid Fossil Deposits

Of all the various types of microalgae within the fossil record, the coccolithophorids are particularly well represented, with continuous records stretching from their first recording in the late Triassic (>210 x 10^6 years/Ma BP) to the present. They, together with an array of other calcareous fragments (nannofossils), also of presumed microalgal origin, dominate chalk deposits covering extensive areas of what were once shallow seas, the result of sedimentation processes similar to those that have occurred over time and continue to occur in today's oceans. Our knowledge and understanding of coccolithophorid fossils has been greatly enhanced through the application of scanning electron microscopy, stimulated by the demands of the oil exploration industry for reliable biostratigraphic information to aid in the dating and characterisation of sediments.

Some of the highlights of the earlier work examining the contribution of coccolithophorid fossils to paleobiological research, including the study of evolutionary processes and environmental change, have been outlined by Bown et al. (1991) and Young et al. (1994). Although calcareous (coccolith) nannofossils appear to dominate the late Triassic fossils, it is probable that some dinoflagellate fossils may also date from about the same time. The apparent extinction of most of the Late Cretaceous coccolithophorid families at or just after the Cretaceous/Tertairy boundary (*ca.*60+ x 10^6 yBP) is interesting especially since neither the diatoms nor the dinoflagellates appear to have suffered a similar fate (Hallam & Perch-Nielsen 1990). It would appear that after this boundary, during the Palaeocene and early Eocene, there followed a period of rapid evolutionary radiation giving rise to most of the modern coccolithophorid families and other related microalgal species, suggesting that modern marine phytoplankton are essentially Mesozoic in origin.

The fossil evidence suggests that the coccolithophorids first appeared as abundant but low diversity assemblages apparently restricted to low latitudes. The high abundances that later became established in the Jurassic and Cretaceous are reflected by the nannofossil-generated carbonate sediments in rocks of those periods. The recovery after the extinction event of the Triassic/Jurassic boundary (*ca.* 210 x 10^6 y BP) was rapid, reaching a second diversity peak in the late Cretaceous. Nannoplanktonic coccolithophorid species, in particular, are considered to have been the major cause of shifts in global calcification from the continental shelves towards deep oceans, with accompanying effects on deep-ocean CO_2 budgets, calcite compensation depths and geological carbonate turnover rates (Bown &Young, 1998).

Dinoflagellate, Diatom and Cyanobacterial Fossil Deposits

The earliest known dinoflagellate fossils are thought to have been formed at about the same time as the coccolithophorid fossils, whereas fossil-forming diatoms and silicoflagellates are thought to have first appeared during the Cretaceous (140-65 x 10^6 y BP). Earlier forms, from the Triassic and even the Precambrian (Sims et al.2006) may have been destroyed through silica re-crystalisation under pressure. Sedimentary records of early photosynthetic microbial communities date from 3-4 billion years bp in the form of surviving laminated and lithified structures (stromatolites) which are dominated by cyanobacterial-like structures, presumed to be the earliest microorganisms capable of oxygenic photosynthesis (Bosak *et al.*2009) and hence enabling the evolution of aerobic microorganisms during the late Precambrian.

The emergence of the diatoms, dinoflagellates and coccolithophorids (in primitive oceans that previously had supported predominantly cyanobacteria and green microalgae) resulted in a major shift in global organic carbon cycling (Armbrust 2009). In particular, the dominance of larger phytoplankton species, with increased sinking rates, would have resulted in enhanced burial of organic carbon in continental margins and shallow seas, creating much of the world's petroleum reserves (Falkowsky et al.2005).

The fossil record for the early Mesozoic ocean sediments shows that the dominant diatom species would have had highly silicified cell walls which would have restricted them to a benthic existence in shallow near-shore waters (Armbrust 2009). Interestingly, the mass extinction (KPB) of the Cretaceous-Paleogene boundary (60+ x 10^6 years BP), which resulted in substantial reduction in the diversity of dinoflagellates and coccolithophorids, left the diatoms relatively unscathed. It is suggested that pelagic microalgal ecosystems might have taken up to ~3 million years to fully recover although ecosystems in the continental margins may have recovered much quicker, possibly as a result of selective survival of benthic cysts or resting stages (Kaiho 1994). Geochemical analysis of the remarkably thick (Fiskeler) clay layer at Kulstirenden, Denmark, has indicated that microalgal primary productivity there may have recovered from the KPB impact rather more rapidly (Sepúlveda et al.2009).

Diatom diversity is believed to have peaked at the Eocene/Oligocene boundary (*ca.* 30 x 10^6 years BP), at a time coinciding with the appearance of raphid, pennate forms with their greatly increased capacity for colonisation of diverse locations. Such raphid forms dominate the diatom communities in many of today's oceans where they are considered to be major contributors to diatom-based export of carbon from surface waters (Zielinski & Gersonde 1997) (discussed in more detail in Chapters II and XI).

The world's richest source of diatomite (diatomaceous earths, mined for a variety of uses) was formed from the remains of diatoms settled in ancient oceans or lakes such as those of the early to middle Miocene (*ca.*20 x !0^6 y BP). Diatomite is used in cleansers, paints, filtering agents and as an abrasive in many products. The most striking historical example of the commercial use of diatomite is its use by Alfred Nobel for the adsorption of nitroglycerine in the production of dynamite. Some of the most complete sections of marine deposits of the Miocene age have been recovered from the Calvert Cliffs on the shores of Chesapeake Bay, USA, the site of vast deposits of two fossil diatoms, *Actinoptychus heliopelta* and *Sceptroneis caduceus* (Gernant et al.1971). Similarly well- preserved remains of diatoms and diatom resting spores of the Oligocene and Miocene periods have been recorded from various other regions of the world (e.g. Suto 2007).

Fossil remains of a range of microalgal species have been extensively used in dating sedimentary and volcanic rocks, especially to confirm information obtained by other means. Diatom fossils and Chrysophycean cysts, in particular, have been shown to be very valuable for dating different strata, certainly in comparison with fossils of vertebrate and invertebrate taxa, which they vastly outnumber in most archaeological sites (Van Landingham 1964, 1991). Chrysophytes generally occur in high numbers, have high rates of growth and relatively rapid rates of evolution making them more paleoecologically indicative and age-diagnostic than larger, less abundant fossils. The diatom and Chrysophycean remains often occur in such abundance as to allow statistically significant correlation of shared taxa between different strata including accurate assessment of the first occurrence of specific taxa and of the relative abundance of different taxa or fossil forms.

A good example is provided by the work of Van Landingham (2010) at the Valsequillo (Hueyatlaco) site, Pueblo, Mexico where analysis of diatom and Chrysophycean cyst remains has contributed to dating of the sediment with sufficient accuracy to draw conclusions regarding the timing of human migration to the Americas. Samples of volcanic rocks at this site have yielded 44 extant and 39 preserved extinct chrysophyte taxa and 467 extant and 78 extinct diatom taxa, many of which are age-diagnostic indicators allowing dating of the Hueyatlaco artefacts to an era between 80 and 400 x 10^3 y BP. All of the indicative diatom/cyst fossils associated with ancient artefacts at Hueyatlaco, are consistent with a Sangamonian Interglacial age or older, enhancing the case for the great antiquity of human presence in the New World.

Oil Shales, Tar Sands and Petroleum

Oil shales are sedimentary rocks similar to the source rocks from which petroleum is obtained but they contain greater amounts (up to 40%) of kerogen, organic matter formed by deposition of silt and organic debris on lake beds and sea bottoms. They may be regarded as occupying one step in the process of decomposition of organic matter to yield such deposits as anthracite, bituminous and brown coal, peat, petroleum and ashphaltum. Beds of oil shale were first described by Alderson (1919) as having been laid down in lakes, lagoons and shallow marine waters where microalgae constituted the major primary source of organic matter. Oil shale beds have a world-wide distribution and the yield of oil can vary from 30-40 gallons t^{-1} to less than half this amount. Different kerogens vary according to their source material and according to the conditions under which they were formed. Some kerogens are almost exclusively of microalgal origin while others are from a mixture of terrestrial and marine sources. Oil shales differ from oil-bearing rocks largely through not having been exposed to sufficiently high temperatures to convert the kerogen to oil. Tar sands are composed of sand, clay and 10-15% organic matter in the form of heavy, black, viscous oil (bitumen). They differ from oil-shales in having their oil content free, lying within pores in the rock.

The source material of various organic fuels have been characterised by tracing their evolution during progressive carbonisation, as expressed by two atomic ratios namely the H:C and the O:C ratios (see Figure 2). It can be seen that as the curves approach the origin (100% carbon), the alginites (organic material of largely microalgal origin) follow a pathway with higher H:C ratio (~ 80%) indicating greater yield of volatilized material from pyrolysis (North 1985).

Of the various forms of alginites, the higher H:C ratios were associated with materials identified as being dominated by *Botryococcus* (resinite) or dinoflagellates, together with a range of other materials (tasmanite) (Jones & Edison, 1978 – reported in North, 1985). Confirmation of the microalgal origin of the organic matter (kerogen) of some sedimentary rocks has come from a combination of microscopy and chemical analyses. It has been shown that the kerogen of the Messel oil shale sediments in Germany, for example, contains cell wall remains virtually indistinguishable from those of an extant microalgal chlorophyte, *Tetraedron minimum* (Goth et al.1988).

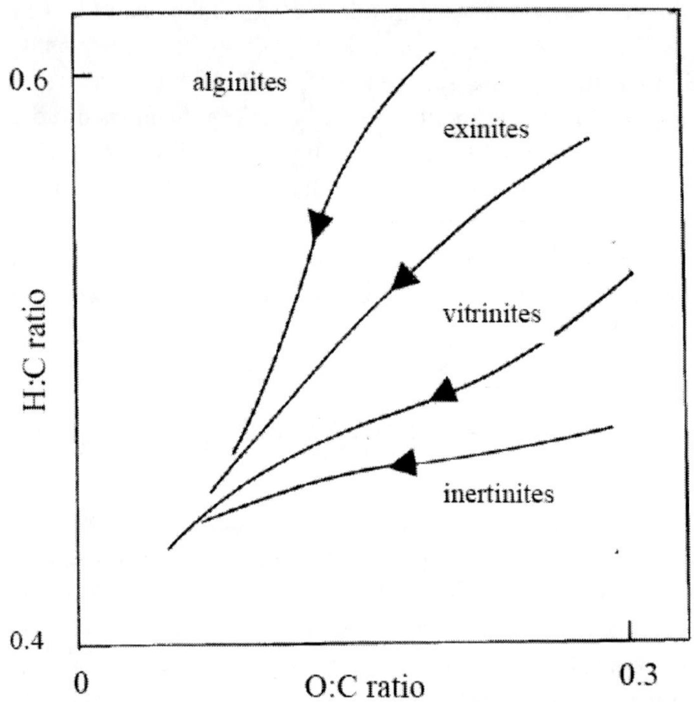

Figure 2. Evolutionary pathways of different organic fuels as expressed by the atomic ratios H:C and O:C: Alginites – principally of microalgal origin, Exinites – mega- and microspore exines, remains of plant resins and waxes, plant cuticles, Vitrinites – woody structures, Inertinites – degradation products of humic acids, lignin and proteins, fungal spore remains. (Redrawn, with permission, from the original data of Van Krevelen (1981) as depicted in Figure 7.4 in North, 1985).

Both the fossil kerogen and present-day *Tetraedron* cells have, moreover, been shown to contain significant quantities of an insoluble, non-hydrolysable highly aliphatic biopolymer material, a likely precursor of the *n*-alkanes in crude oils. Different types of kerogen (classified according to C, H and O composition) are formed under different conditions and from different microalgal or plant cellular components.). Kerogen from microalgal lipid material deposited under anoxic conditions in lake or marine sediments is recognised as one of the world's richest source of petroleum. Sapropels are formed primarily from carbonaceous material of aquatic origin such as the kerosene shales derived almost exclusively from the green microalga *Botryococcus braunii* (Chlorophyta). Other forms of sedimentary material of microalgal origin are the shungites, various bitumens and diatomites, the latter composed of the siliceous frustules of diatoms and their associated organic components (Combaz 1980).

Botryococcus is of particular interest because it has been shown to be associated with the formation of a variety of oil-rich geological deposits dating from the Ordivician period (see review by Cane 1977). The significance of this microalga in such deposits is indicated by the presence of botryococcane, a hydrogenated derivative of botryococcene ($C_{30}H_{50}$), which has been shown to occur at concentrations of between 0.9 and 1.4 % in two Sumatran crude oils (Moldowan & Seifert 1980). Other oil shales known to be of microalgal origin include the Torbanites and the Tasmanites. The former have been shown to contain remains of *Botryococcus* and are deemed to be synonymous with Kukersite, a microalgal shale of Ordivician age that occurs in Estonia. The principal organic component of Kukersite is telalginite, derived from the cyanobacterium *Gloeocapsamorpha prisca* (Dyni 2003). Modern deposits analogous with the Torbanites have been described from various locations. Those from the Coorong lagoon in South Austrailia, for example, occur as gelatinous yellow-green masses which dry to a dark rubbery material whose microalgal composition is evident from microscopic examination. Tasmanites were first recognised in Tasmania and in the Brooks Range of North Alaska but their occurrence and distribution are widespread over the world, principally in lake areas, fluvial dominated deltas and certain marine shelves. Oil shales from the Green River Formation of the Central United States have been shown to contain ca. 11% kerogen and 3% bitumen (Yen & Chilingarian 1976). The recovery and extraction of useable oil products from these sources requires the application of specialised technologies. Advances during recent decades in hydraulic fracturing (fracking) techniques have allowed access to vast deposits of crude oil trapped between layers of shale sediments and it is anticipated that improvements in extraction techniques (combined with reduced yields from current conventional sources) will undoubtedly make kerogen-rich sources such as those of the Green River Formation (and similar deposits elsewhere) commercially attractive.

Kerogens from a range of marine and lacustrine source rocks, previously regarded as amorphous, are now recognised as being composed of very thin (10-30 nm thick) lamellar structures which on analysis, have been shown to contain non-hydrolysable material resembling that in the outer walls of extant microalgae. Chemical analysis of the kerogen ultralaminae and comparison with pyrolysis products of wall material from the microalga *Chlorella fusca,* for example, have confirmed that the fossil ultralaminae are derived from material (algaenans) from microalgal cell walls (Derenne et al.1992). It was further concluded that differences between the 2-n-alkanones generated on pyrolysis of different kerogens may reflect differences in the populations of the source microalgae.

The contribution of coccolithophorid blooms to sediments that formed over geological times in coastal regions of Eastern Europe is well illustrated by rock cores taken near Kimmeridge Bay, Dorset and from other regions of southern England (Gallois 1976). The thickness of the so-called Kimmeridge Clay oil-shale bands in those rocks varies over the area from <100m to >500m with the coccolithophorid content ranging from 60 to 98 % of the calcium carbonate fraction of the rock. The coccolithophorid-rich bands are composed almost entirely of one fossil species *Ellipsogelosphaera britannica* indicating that it represents the remains of a microalgal bloom dominated by one or a few species. Gallois (1976) concluded that the Kimmeridge Clay oil shales (and hence much of North Sea oil) formed from microalgal blooms in an enclosed basin environment separated from the open ocean. The anoxic bottom conditions beneath a body of undisturbed, stagnant sea water, and the probable poisoning effects from some other microalgae such as dinoflagellates, would have created conditions suitable for the formation of rich organic sediments. The preservation in such oil shale of a substantial amount of organic matter (kerogen) is probably attributable to high microalgal productivity in shallow, continental shelf, nutrient-rich, and largely anaerobic waters. It has been estimated that over 70% of the total hydrocarbon (oil or oil equivalent) of the North Sea is derived from Jurrassic source rocks (Spencer et al.1996) with the Brent and Kimmeridgian kerogens as major sources.

Measurement of nitrogen isotopic ratios ($\delta^{15}N$) in late Pleistocene chlorophyll of organic-rich sapropels from the eastern Mdeiterranean Sea has shown that, contrary to to-day's well ventilated, nutrient depleted, low productivity conditions, those of that period were very different (Sachs & Repeta 1999). They allowed formation of green-brown to black deposits with organic carbon concentrations of 2 to 5% interspersed between grey nannofossil and foraminiferal marl oozes of lower carbon concentrations, to yield layers 1 to 30 cm thick that formed over periods of 1,000 to 10,000 years. The $\delta^{15}N$ data were interpreted by Sachs & Repeta (1999) to indicate a pattern of sapropel formation in which increased river discharge facilitated development of a specialised phytoplankton population whose annual mass sinking provided the organic flux to generate the sapropels.

Chapter II

Microalgae and Global Climate

Microalgae have been extensively used as indicators of past climate changes. Fossil microalgal diversity records have been used as indicators of climate change, at least over the Cenozoic era (66 Ma to present). Good correlation has been reported between diversity and climate trends, with higher diversities generally associated with warm climates (Bown & Young 1998). The Cenozoic diversity maximum for coccolithophorid fossils (and total nannofossils) coincides with the Paleocene/Eocene thermal maximum while the climate cooling through the Late Eocene and into the Oligocene is tracked by a significant decline in species diversity.

Diatoms preserved in lake sediments have been used as a basis for climate change reconstructions on both geological time scales as well as those covering more recent periods. They contribute to such reconstructions by providing baseline data against which any short-term variability can be assessed. Models defining diatom-climate linkages have generally been more problematical than those applied to other environmental factors such as salinity, pH and nutrient changes. Anderson (2000), in his review of relationships between diatom functions and climate noted that of the various factors influencing diatom ecology, temperature is probably much less important than some other factors such as water chemistry, turbulence and thermal stratification. Moreover, direct climatic effects are unlikely to be independent of other factors and, over long time periods especially, effects on soil and vegetation will tend to dominate lake-catchment hydrochemistry and hence lake biota. In such cases, diatom-temperature models may reveal inferred temperatures that may merely be artefacts of changing lake-water chemistry.

Lake-Sediment Microalgae As Indicators of Climatic Changes

Siliceous microalgal fossils preserved in lake sediments can provide evidence of climatic changes over time within the catchment. Trends in population succession are normally interpretable on geological time scales such as those described by Bradbury (1988) who noted that the increase in *Aulacoseira* sp. in Elk Lake (Minnesota, USA) during mid-Holocene (5400-4800y BP) was a reflection of the increased mixing of the water column associated with seasonally strong late-summer thunder storms. Examination of the diatom assemblages distributed along a 280 cm-long core taken from Liyu Lake (eastern Taiwan), allowed Wang et al. (2009) to identify three different habitat types that the area had passed through over the period 1760 BP(250 A.D) to the present. The diatom flora of the period before 250 A.D. (dominated by *Fragilaria construens*) suggested a riverine habitat. A subsequent change in the diatom assemblage, with a marked increase in the abundance of *Cyclotella stelligera*, was interpreted as being indicative of increased precipitation and consequent lake formation. Two peaks of significant changes (1370 and 1690-1780 AD) were also identified and interpreted as indicating severe drying events. The effects of more recent global warming and an increase in the intensity of Eastern Asia Monsoons were indicated by a higher abundance of planktonic species in the diatom assemblages.

Studies of Great Lake sediments have provided evidence of ecological changes that have taken place over much smaller time scales, sometimes no more than a few decades. In their study of the recent history of Lake Ontario, Stoermer et al. (1985) noted striking modifications of the siliceous microalgal fossils (diatoms and chrysophytes) in radiometrically dated sediment cores. Over the time period investigated (1700 – 1979), they report evidence of local extinction of some taxa (e.g. *Fragilaria intermedia* var. *fallax*, *Melosira italica* and some *Cyclotella* species) and the apparent introduction of other taxa which later became abundant (e.g. *Stephanodiscus parvus*, *Asterionella formosa* and *Diatoma tenue*). Without exception, the populations displaced were those normally abundant in lakes receiving low to moderate nutrient loadings. Moreover, the replacement taxa were those characteristic of lakes receiving high to very high nutrient loads. Stoermer et al. (1985) conclude that the major factor driving species succession in Lake Ontario was phosphorus enrichment (due to human influence) and accompanying depletion of available silica supplies. Substantial changes in the primary production base, they noted, occurred well in advance of similar changes at higher trophic levels, reflecting similar patterns observed in other Great Lakes (Beeton 1969).

It was noted, however, that the increased production inferred from qualitative changes in microalgal populations did not appear to be reflected in a steadily increasing trend in siliceous microfossil abundance. Rather, the highest absolute numbers of siliceous microfossils occurred in the 1859 sample with a general trend of decline in abundance both below and above that sample depth. As a possible explanation for this, it was suggested that the expected post-1860 trend was, in fact, masked by increasing allochthonous sedimentation. Other factors that, according to the authors, have to be taken into consideration in interpreting sediment microfossil data are those relating to seasonality and to the availability of a physical niche for survival of a particular species.

Over a much longer time-scale, diatom remains in sediments from Lagoa Grande (northwest Spain) dating from ca. 9800 BP have been reported to exhibit successive marked changes at 8200, 6000, 4200 and 2600 yr. BP (Leira 2005), which have been interpreted as responses to climatic changes identified from marine and terrestrial palaeoclimatic data. During the earlier period, the lake was deemed to be shallow with the diatom population dominated by *Aulacoseira* species characteristic of acidic and oligotrophic waters. Then, the gradual appearance of *Aulacoseira ambigua* in the early Holocene was interpreted to indicate an increase in water depth and in nutrient levels. The observed subsequent changes in the diatom assemblages throughout the Holocene were linked to fluctuating limnological conditions, coinciding, in turn, with regional, climate-related vegetation succession. Diatom responses over the final 2000 year period, reflected in progressive peatland development and growth of littoral vegetation, point to catchment processes as being major controlling factors in the characteristics of this lake.

Oceanic Phytoplankton and Climate

Development of satellite-based colour scanners and associated spectrometers and image spectro-radiometers has provided valuable information on the interaction of microalgae and the various physical processes acting on oceanic surface waters. The work of d'Ovidio et al. (2010), for example, has allowed partitioning of bulk optical signals into major phytoplankton functional groups. They found that oceanic phytoplankton populations are organised into (sub)-mesoscale patches (10-100km) of dominant types separated by physical fronts induced by horizontal stirring. Microalgal groups that were identified include diatoms, coccolithophores, nanoeukaryotes and cyanobacteria (including nitrogen-fixing species such as *Trichodesmium*). This level of precision allows identification of fluid dynamic niches where phytoplankton assemblages interact with a distinct set of physicochemical parameters. As reported by Kudela (2010), some phytoplankton groups such as diatoms have been confirmed as contributing disproportionately to fisheries food webs and to the export of organic matter to oceanic depths. Similarly, coccolithophorids have a strong influence on the carbon chemistry of surface oceanic waters through the formation of calcium carbonate shells. The chrysophyte, *Phaeocystis,* has been reported to produce dimethyl sulphide, whose molecules, by acting as cloud condensing nuclei form sulphate aerosols, thus significantly affecting the heat balance of oceanic surface waters (see Chapter III).

Observational microalgal data can now be coupled with physical data to provide information on the forces driving the biogeochemistry of surface oceanic waters. Regions of strong horizontal stirring have, for example, been identified as areas of high biological diversity, allowing a range of phytoplankton types to thrive (Barton et al. 2010; d'Ovidio et al. 2010). Considerable further refinement of these techniques would, however, be required before the ecological landscape of oceanic waters can be adequately mapped at spacial and temporal scales sufficient to predict any biogeochemical changes that might occur in the future.

On a broader scale, modelling of climatic events originating in the north pacific subtropical gyre - a massive region of circulating oceanic currents - has revealed a strong correlation between the number of storm events which form over that region and the concentration of surface phytoplankton. Models produced by Gnanadesikan & Anderson (2009) have predicted that if the region contained no absorbing and scattering materials (i.e. very low phytoplankton concentrations), subtropical cyclone activity would be reduced by about two-thirds. Cyclone tracks, instead, would be concentrated along the equator. Halving existing phytoplankton standing stock, it was predicted, would reduce the frequency of storms by 35%.

It was suggested that greater phytoplankton numbers would reduce heat penetration down the water column, causing greater surface warming, whereas at low phytoplankton concentrations, heat is more effectively carried away by deep currents. The consequent cooling would changes the pattern of air convection in the region so reducing the frequency of cyclones. Such an interpretation would be consistent with historical records from the region which show a period of lower than usual cyclone activity during the 1960s when chlorophyll concentrations were 50% lower than more recent values.

The key role that diatoms and dinoflagellates play in ecosystem processes in NE Atlantic waters has been described by McQuatters-Gollop et al. (2007). Based on >50 years of data, they were able to confirm that these two phytoplankton groups display dissimilar bloom patterns with the diatom spring bloom peaking during May before gradually declining through mid-summer and then wekly peaking again in late summer. Dinoflagellates bloom strongly during late summer, peaking in August, before progressively declining throughout autumn. Both groups showed minimum abundances during winter months. Such information is invaluable as baseline data against which to assess possible future effects of climate change and changing anthropogenic pressures.

Phytoplankton and Climate Predictions

For marine microalgae, as for plankton in general, it has been said that their biogeographical distribution is primarily a function of the geographical development of the ecological habitat to which the different taxa are adapted (Burnett et al. 2000). Thus, among modern marine microalgae, there are no obvious examples of species whose biogeographical range is restricted by dispersal limitation. This has allowed nannofossil palaeobiogeographical patterns to be interpreted as being primarily records of marine climate change

The world oceanic fossil record of the three major armoured eukaryotic phytoplankton taxa, the diatoms, the coccolithophores and the dinoflagellates, has been interpreted to reflect gross climatic changes that might have occurred over the past 200 x 10^6 years (200 Ma) (Falkowski & Oliver 2007). Because these three phytoplankton groups occupy different mixing regimes in the modern ocean, it can be presumed that their fossil record in marine sediments probably reflects previous environmental conditions, in particular those relating to geochemical parameters for ocean mixing. From the early Triassic period until the extinction event that occurred 65 Ma ago, the diversity of the carbonate-precipitating coccolithophores and armoured dinoflagellates increased, with an accompanying increase in the export of organic matter to marine sediments, although oceanic productivity was probably low. Fossilized diatoms were first recorded at approximately 146 Ma (Koistra et al. 2007) but were rare throughout the Mesozoic period. During the Palaeogene period, with a significant portion of the coccolithophores and dinoflagellates already removed by the extinction event, there was a rapid increase in diatom diversity. This reflects what must have been a major climatic shift and a substantial change in oceanic thermal structure leading to increased export of organic carbon to continental margins (Smetacek 1999).

The climatic changes inferred from the observed changes in the phytoplankton fossil record (in particular the increased diversity and decreased average size of the diatoms during the Cenozoic Era, as described above), have been used to predict possible future climatic changes. The continued accumulation of greenhouse gases is likely to lead to increasing isolation of the upper mixed layer of the ocean from the deeper waters. (Levitus et al. 2000). Over most of the open ocean, this increased stratification will reduce nutrient availability so reducing the competitiveness of the diatoms. The coccolithphorids and other smaller phytoplankton (including the cyanobacteria and *Prochlorococcus*), on the other hand, will become more dominant (Falkowski & Oliver 2007). How these changes at the primary producer level will affect higher levels of the marine food webs and global biogeochemical fluxes, is more difficult to predict.

Microalgae and Increasing Atmospheric CO_2 Concentrations

Long-term records have indicated a 40% increase in surface sea-water CO_2 concentrations since the industrial revolution (Caldeira & Wickett 2003), with estimations that the rate of increase has tripled since the mid-20th century (Fussel 2009). Coastal zones, which receive additional CO_2 from up-welling (Feely et al. 2008) or terestrial run-off (Cai et al. 2011) have generally experienced even larger increases. It has been estimated that continued anthropogenic carbon emissions at the present rate would probably result in a doubling of the world's atmospheric CO_2 concentrations by the end of the present century, with an approximately 12% increase of total inorganic carbon (TCO_2) in surface oceanic waters compared with pre-industrial levels (Houghton et al. 2001).

Other predicted consequences will be a drop in average surface water pH from *ca.* 8.2 to *ca.* 7.8 with accompanying changes in equilibrium pools of TCO_2 species (i.e. free CO_2, HCO_3^- and CO_3^{2-}). These changes have the potential to impact strongly on the microalgae, especially on their photosynthetic productivity. At prevailing surface seawater pHs, microalgal photosynthesis is limited by the rate at which CO_2 is fixed by the enzyme ribulose bisphosphate carboxylate (RUBISCO) *via* the C_3 photosynthetic pathway. Any decrease in surface water pH, in addition to increasing the concentration of free CO_2 and so increasing the rate of CO_2 fixation, may also affect a range of intracellular processes some, no doubt, with negative effects on phytoplankton growth. Other effects, difficult to predict, may result from altered species composition within the phytoplankton community. The effects of surface water acidification might be expected to be particularly noticeable on calcifying microalgae (such as the ecologically important coccolithophorids) (Riebesell 2004) through effects on calcite and aragonite saturation states (Hurd et al.2009). Any oceanic acidification that might occur would also have effects higher up the food chain, such as on bivalves, through effects on calcification (Doney et al. 2009) and hence, indirectly, on the microalgal population.

Seawater, because of its high content of dissolved inorganic carbon (DIC), is generally well buffered, with a stable pH at around 8.2. In some locations, however, such as nutrient enriched fjords and lagoons, high levels of microalgal photosynthetic CO_2 uptake may exceed the rate of its replacement from the atmosphere, resulting in pH values greater than 9.5. Such high pH values and low DIC concentrations would be expected to limit phytoplankton photosynthesis and to favour those species able to utilize HCO_3^- (at greater energy cost). In those situations, an increase in atmospheric CO_2 concentrations would promote phytoplankton productivity and perhaps increase the frequency or magnitude of blooms. Some microalgal species, including species of *Ceratium* that dominate the flora of some coastal waters, are known to be very sensitive to high pH (Schmidt & Hansen 2001), with upper limits of growth as low as 8.3 or 8.4. For three species of *Ceratium* (*C. furca, C. tripos,* and *C. fusus*) studied by Søderberg & Hansen (2007) the upper pH limits for growth were considerably higher (between 8.7 and 9.1), but still well below the maximum pHs periodically recorded in some coastal waters. The sensitivity of some *Ceratium* species to high pHs was found to be due to a combination of DIC limitation and direct pH effects. Berge et al. (2010) tested the effects of lowered pH (by HCl addition) on the growth rates of eight common species of phytoplankton. They found that for the dinoflagellate *Heterocapsa triquetra* and the cryptophyte *Teleaulax amphioxeia*, the lower pH limits for growth were pH~6.0 and 6.3 respectively and growth rates of these two species were significantly reduced at pHs within the range pH 6.4 to 6.5. Within the pH range 7.0 to 8.5, however, none of the phytoplankton species tested (representing diatoms, dinoflagellates, cryptophytes and haptophytes) showed any increased or decreased growth and production rates in response to pH changes, perhaps because they have CO_2 concentrating mechanisms which render them insensitive to increasing concentrations of free CO_2. It would seem, therefore, that marine phytoplankton in general would be resistant to surface water acidification, certainly within the range anticipated from predicted increased atmospheric CO_2 concentrations. Full strength oceanic waters (with high TCO_2 concentrations of 2.2-2.4 mM) would be well buffered against pH changes. Truly oceanic species, however, would be less likely to be acclimatised to the kind of pH fluctuations which are common in coastal surface waters on both daily and seasonal time scales. The sensitivity of two of the species to the lower pHs was interpreted by Berge et al. (2010) as being primarily due to direct pH effects on cellular processes rather than changes in TCO_2 and the various carbon equilibrium pools.

Two phytoplankton groups in particular, the diatoms and coccolithophorids, play a key role in the regulation of CO_2 exchanges between the atmosphere and the upper oceanic waters and in the export of carbon (as organic matter) into the ocean depths (the "biological pump"). In the modern ocean, diatoms dominate the export flux of carbon, contributing significantly to the long-term sequestration of atmospheric CO_2 in the ocean interior (Smetacek 1999). Coccolithophorids, through the formation of calcium carbonate ($CaCO_3$) shells, influence the inorganic carbon equilibria and alkalinity of sea water with the potential to release CO_2 which may escape to the atmosphere (Frankignoulle et al. 1994) (the "alkalinity pump"). The interaction between microalgal calcification and sea water is chemically complex but can be summarised by the single equation:

$$Ca^{2+} + 2HCO_3 = CaCO_3 + CO_2 + H_2O \text{ (Berger 1982)}$$

which shows that precipitation of calcium carbonate results in an increase in the concentration of dissolved CO_2. On a global biogeochemical scale, limestone formation by coccolithophores (and other calcifying biota) has been identified as a probable significant factor affecting atmospheric CO_2 concentrations and hence global warming (Holligan & Robertson 1996). The balance between the coccolithophores and diatoms will, therefore, affect the balance between $CaCO_3$ export and the export of organic carbon, a major factor determining the partitioning of CO_2 between the ocean and the atmosphere. Prior to the evolution of nannoplankton in the late Triassic (220 Ma), there was probably little production of pelagic carbonates. Then, following the coccolithophore-related shift toward pelagic carbonate deposition, there would have been an increased metamorphic recycling of carbonate (summarised as $CaCO_3 + SiO_2 = CaSiO_3 + CO_2$) (Volk 1989) and, in the long term, a consequent increase in global CO_2 levels.

Cermeño et al. (2008) investigated phytoplankton biomass and species composition in relation to a range of hydrographic variables along four Atlantic Meridional Transects extending from latitude 60°S to 60°N. They looked specifically at the relative abundance of diatoms and coccolithophorids with particular reference to the "biological pump" and the "alkalinity pump" mentioned above. They showed that diatom biomass increased more than five orders of magnitude from low latitudes to temperate and upwelling systems, whereas coccolithophorid standing stocks varied by only two orders of magnitude. The variations in diatom and coccolithophorid biomass were further reflected in their patterns of diversity. The number of diatom species, in particular, increased rapidly from subtropical to temperate and upwelling regions, whereas the coccolithophorids exhibited comparatively minor responses to environmental variables. The balance between diatoms and coccolithophorids, primarily driven by variations in diatom biomass and diversity, bore a striking correspondence with the nutricline depth, an indicator of nutrient supply to the upper mixed layer of the ocean. Based on these observations, Cermeño et al. (2008) conclude that increased thermal stratification predicted to occur over the coming century would result in a dramatic reduction in nutrient supply to the euphotic zone. This, in turn, would alter the composition of the phytoplankton community with a consequent decreased efficiency of the diatom-driven biological pump sequestering atmospheric CO_2, resulting in a positive feedback acting across the oceanic-atmosphere interface.

It has been estimated that large-size classes of phytoplankton such as the diatoms tend to dominate phytoplankton communities in well-mixed coastal and upwelling regions (Armbrust 2009). They also dominate along the sea-ice edge where sufficient light, inorganic nitrogen, phosphorus, silicon and trace elements are available to sustain their growth (Morel & Price 2003). In the much larger oligotrophic regions of the oceans, the photic zone tends to be dominated by small-sized picoeukaryotes, cyanobacteria and prochlorophytes. All of these primary producers contribute to the so-called organic pump which involves conversion of atmospheric CO_2 to microalgal biomass. Much of the fixed CO_2 is, however, re-released to the atmosphere through various routes (e.g. metazoan grazing, microbial/viral degredation and carbonate production by calcifying organisms) but a small proportion, estimated as between 5 and 20%, will sink to deeper waters where it remains isolated from further interaction with the atmosphere unless returned to the surface by deep ocean currents and upwelling processes (Denman 2008). It is estimated that only a very small amount of planktonic debris (0.1%) ever reaches the ocean sediments and remains buried there (Middleburg & Meysman 2007).

Given the major contribution of cyanobacteria (*Synechococcus, Trichodesmium* and *Prochlorococcus*) to open-ocean phytoplankton, especially in the subtropical and tropical gyres (e.g. Partensky et al. 1999; Capone et al. 2005), particular attention has been directed to the response of these species to ongoing acidification of surface waters. Studies with cultures of *Trichodesmium erythreum* have shown that under nutrient replete conditions, increasing pCO_2 (increasing acidity) consistently resulted in a substantial increase in both N_2 fixation (35-120%) and C fixation (15-100%) due to alleviation of inorganic C limitation at low ambient pCO_2 levels (Hutchins et al. 2007). These increases resulted in increased production of dissolved organic matter which, if replicated in natural populations, would suggest that ocean acidification could have far-reaching effects on the role of this cyanobacterium in oceanic carbon and nitrogen cycles. Natural populations dominated by *Trichodesmium* were found by Lomas et al. (2012) to respond similarly to changes in pH/pCO_2 when tested over the range ~8.4/~150ppm. to ~7.8/~800ppm. Populations dominated by *Prochlorococcus* and

Synechococcus showed no such clear response of C-fixation rates to changes in pH/pCO_2. Instead, there was evidence in these species of rapid acclimation of cellular physiology to new pH/pCO_2 conditions. Laboratory studies with the marine cyanobacterium *Crocosphaera* have shown that rates of N_2 and CO_2 fixation (and growth) increased in step with increasing partial pressure of CO_2 but only under Fe-replete conditions. (Fu et al. 2008). Similar responses, if repeated in natural cyanobacterial populations, would mean that anthropogenic CO_2 enrichment, although tempered by Fe availability, could lead to elevated inputs of new nitrogen to the ocean and possible consequent Fe and/or phosphorus limitation. Clearly, various components of natural phytoplankton populations respond differently to increasing pCO_2, making it very difficult to predict possible future trends.

The concentration of CO_2 in present day sea water is well below the K_m [CO_2] for the RUBISCO enzyme system responsible for CO_2 fixation (Badger et al. 1998) but marine diatoms, and indeed many other marine microalgae, posess active uptake and accumulation mechanisms for dissolved inorganic carbon to support their photosynthesis. Such carbon concentrating mechanisms (CCM) involve active uptake of both CO_2 and HCO_3^- (Johnston et al. 2001; Matsuda et al. 2001). CCM activity has been shown to be suppressed when CO_2 concentrations are high and induced under CO_2-limiting conditions. Such regulatory mechanisms are clearly an important component of the complex responses to increased CO_2 concentrations at the ocean surface

The complex interactions between pCO_2, phytoplankton, heterotrophic bacteria and extracellular enzyme activities have been investigated using mesocosm culture vessels in which CO_2 concentrations are adjusted to represent glacial (190μatm), present day (370μatm) and future (750μatm) conditions. Nutrients were added to each mesocosm to induce phytoplankton blooms and the responses of bacteria, phytoplankton and zooplankton to the different pCO_2/pH levels noted. Allgaier et al. (2008) observed that changes in the make up and physiological state of the phytoplankton in response to changed pCO_2 yielded qualitatively different dissolved substrates and consequent changes in the heterotrophic microbe community (Grossart et al. 2006). Others have suggested that these effects of pCO_2 through their effects on the initial steps of heterotrophic carbon cycling would strongly impact the microbial loop and the recycling of organic matter in oceanic surface waters (Arnosti et al. 2011). The complexity of the various components of the microbial loop makes predictions of future likely responses to pCO_2 very difficult.

Microalgae and Global Temperature Changes

Global climate changes monitored over the last 40 or so years point to a gradual increase in average temperatures, a trend that is predicted to continue into the future with possible significant effects on a range of different ecosystems. Warming effects on the world's oceans, as comprehensively reported by Barnett et al. (2005), impinge directly upon the phytoplankton, and on their special role in the cycling of carbon. Among the effects that might be expected to arise as a consequence of climate changes would be changes in the microalgal species composition (including possible species extinction) with consequent flow-on effects in the food chain and on the export of carbon from the euphotic zone to the lower depths. Observed thinning of the ice cover in the Chukchi Sea, Alaska region (Arrigo, 2012) is thought to be linked to the appearance of phytoplankton blooms, in some cases extending down for 70 metres. The predicted earlier appearance in the year of such blooms might be expected to have a major impact on the entire Arctic food web and on the fisheries.

The effects of sea-surface warming on the marine biota and the resulting impacts on marine carbon cycling will differ depending on the prevailing light and nutrient conditions in the upper layer and on the composition of the pelagic community. Beaugrand et al. (2010) attempted to predict the potential impact of climate change on key marine planktonic groups as monitored by the Continuous Plankton Recorder (CPR) deployed in the North Atlantic over the period 1960-2005. Three planktonic groups were studied: diatoms (identified to the level of genera), the dinoflagellate genus *Ceratium* and calanoid copepods (representing microalgal grazers and identified to species level). They noted that there had been a pronounced latitudinal increase in phytoplanktonic and zooplanktonic diversity in the extratropical North Atlantic Ocean over recent decades and that these changes were at least partially temperature-related. The results further indicated that increasing maximum sea surface temperature (SST) may favour biodiversity of both copepods and *Ceratium* and that decreasing annual variability in SST may contribute to reduction of biodiversity and abundance of diatoms (major contributors to carbon export in these waters). The rise in biodiversity of the copepods was paralleled by a decrease in the mean body size reflecting increased dominance of the population by smaller organisms, with consequent effects on the flow of carbon through the food chain.

Based on their analysis of the CPR data, Beaugrand et al. (2010) predicted that increasing SST in the North Atlantic Ocean and the resulting stronger stratification, may lead to a shift in the phytoplankton community from diatoms (major exporters of carbon to depth) to coccolithophorids. Stronger stratification and the consequent reduced nutrient input into the euphotic zone would affect efficiency of the carbon pump with organic carbon having a longer residence time in surface waters where it would be processed through smaller-sized zooplankton and thus dissipated through more complex food webs. Negative effects of increasing SST on biota at the upper end of the food chain were also predicted, as exemplified by stocks of Altantic cod (*Gadus morhua*) which were strongly influenced by climate–driven effects, especially at the edges of their distributional ranges.

Global warming effects similar to those dscribed above for oceanic phytoplankton populations have also been observed at lake-ecosystem level (Mc Knight et al. 1996; Livingstone & Dokulil, 2001; Magnusson et al. 2003). The model devised for predicting likely responses of phytoplankton to environmental change has indicated that, for the eight phytoplankton species studied, the response to increasing temperature was reduced at low nutrient levels (Elliott et al. 2006). Similarly, the model indicates that, for a shallow lake, the effect of climate warming on the intensity of the spring bloom of diatoms was more marked under hypertrophic conditions (Huber et al. 2008). The link between the trophic status of a lake (Lake Geneva, Switzerland and France) and the respone of its phytoplankton population to warming was further studied by Tadonléké (2010). Using data collected over a 31-year period, it was shown that the effect of warming upon maximum phytoplankton photosynthetic productivity (normalized on a chl *a* basis) was positive and constant when total phosphorus (TP) in the lake water exceeded 22μg PL^{-1} (i.e. eutrophic conditions) but negative when TP was lower (nutrient poor conditions). The significance of these findings and their implications in relation to predicted warming trends will require careful analysis of the complex and nonlinear ecosystem dynamics involved, of which the nutrient status of the lake is only one, albeit a very important component.

Other factors inevitably affected by any increase in sea surface temperatures (e.g. in the higher latitudes of the northern hemisphere, as reported by Belkin (2009), would be those acting through the increased activity of prokaryotic heterotrophs (Apple et al. 2006). Other effects suggested, but yet to be unequivocally confirmed (Feng et al. 2009), might act through the level of viral infection of micralgae. It is likely, however, that different oceanic regions will respond differently to changes in surface temperatures. Thus, in cold water systems, there is a trend for viral infections to increase with increasing temperatures (Danovaro et al. 2010). In some oceanic waters, there is little doubt that increased stratification (due to surface warming) can cause severe nutrient limitation which, in some but not all cases, can result in increased viral infection of the microalgal component with consequenr effects on carbon cycling.

Microalgae As Indicators of Global Acidification

Concerns over increasing acidification of the atmosphere, and of aquatic ecosystems, from industrial contamination over the last forty or so years has stimulated search for reliable indicators of the extent and nature of such acidification trends. In particular, the need for reliable historical records of acidification has focussed on the diatoms in lake sediments, long used for reconstructing pH changes (Battarbee et al. 1986). The techniques used in such reconstructions have been reviewed by Battarbee & Charles (1987) and Birks et al. (1990). They confirm that diatom populations are strongly related to pH or to factors that covary with pH such as alkalinity. When supported by culture experiments they can be interpreted to explain the loss of certain taxa as the pH drops below a critical value. The diatom technique has been used to reconstruct the pH history of a lake (and hence the atmosphere), both in terms of trends and of past pH values. Lakes in sensitive bedrock areas within regions of high acid deposition have been shown to suffer acidification whereas similar sites outside these regions have not. Sensitive sites located close to regions of early industrial development began to be acidified before sites in countries where industrialisation is a more recent occurrence (Battarbee & Charles, 1987).

The comprehensive review by Davis (1987) has described how analysis of the diatom remains and of the resistant scales of various chrysophytes (i.e. golden-brown microalgae) in various lake sediment cores, has contributed to the interpretation of the effects of acidification. Certain clear water lakes, (with pHs between 4.3 and 5.0) were shown to have attained these low pHs within recent decades although they were already acidic (pH 5.0-6.4) before this. One of the consequences of acidification may have been a marked decrease in humics, reducing the availability of organic ligands for detoxification of metal ions. Diatom analyses of sediment cores indicated long-term lake acidification in Late-glacial and Holocene periods but at a much slower rate than the more recent, anthropogenically-related acidification (Battarbee & Charles 1987). Similarly, Birks et al. (1990), in their reconstructions of pH changes in stratigraphic samples from Round Loch of Glenhead, Scotland covering the past 10,000 years, reported only slight fluctuations in pH from late glacial (10,000 BP) until about 1900, but significantly more rapid pH change per unit sediment depth between 1874 and 1931, the time of increased acid deposition.

Diatom-based pH transfer functions have been applied to a large number of fossil records in Europe and North America. The reconstructions showing the extent and timing of acidification, when used in combination with other palaeoecological analyses, have generally identified acidic precipitation (rather than afforestation practices) as the major causal factor (Cumming et al. 1992). In direct contrast to the studies listed above, there are others citing instances where lakes exposed to industrial or mining impacts that might have been expected to cause strong acidification have, in fact, shown no such trends. Thus, studies of sediment cores from eight acid-sensitive lakes in a region of northeastern Alberta, Canada influenced by industries based on bitumen extraction, showed no evidence of acidification (Hazewinkel et al. 2008). Instead, many of the lakes, despite their exposure to enhanced atmospheric SO_2 and and NO_x emissions, showed characteristic changes towards greater productivity and, occasionally, greater alkalinity. The changed responses in these lakes, it was suggested, were probably driven by a complex of interactions involving in-lake production, internal nutrient loading and climate change.

In oceanic waters, acidification of surface layers, either from anthropogenic causes (Orr et al. 2005) or from other causes such as CO_2 production during pelagic biogenic calcification (Merico et al. 2006), has the potential to impact strongly on oceanic carbon cycles. Increased precipitation of acid rain may also come from direct microalgal sources such as the widely occurring dimethylsulfide-producing blooms of *Phaeocystis* (and other marine phytoplankton) (Liss et al. 1992). The dominant contribution to CO_2 cycles by calcifying marine microalgae such as the coccolithophores is well known (Milliman, 1993) as also is the role of these microalgae in facilitating the transfer of organic carbon from surface waters to oceanic depths (Zondervan, 2007). The world-wide distribution of coccolithophore communities from tropical to subarctic and arctic regions (Winter, 1994: Charalampopoulou et al. 2011) ensures they would be strongly impacted by acidification (or global warming) usually manifest as changes in species composition.

Microalgae and Solar UV-B Radiation

Depletion of the stratospheric ozone layer by atmospheric pollutants such as chlorofluorocarbons (CFCs) has resulted in increasing incidence of solar UV-B (280-320 nm) on the Earth's surface (Curtzan, 1992; Madronich et al. 1998). Besides exerting adverse effects on terrestrial vegetation with inevitable impacts on mankind and the entire global ecosystem, UV-B is also known to penetrate into marine and freshwater bodies (Smith et al. 1992). Effects on aquatic organisms are therefore to be expected and, since about 50% of the earth's primary production has been estimated to be attributable to aquatic ecosystems (Siegenthaler & Sarmiento, 1993) attention has been directed to considerations of the sensitivity of microalgae to UV-B exposure and, in particular, on any protective mechanisms they possess (or may be able to develop) to withstand such exposure. In addition to the deleterious effects of UV radiation, Gao et al. (2007) have also reported some positive effects such as is noted in tropical marine phytoplankton where UV can act as an additional source of energy for photosynthesis.

A range of deleterious effects of enhanced UV-B exposure has been reported for many microalgal species such as *Chlamydomonas reinhardtii,* where exposure results in the production of mutants with defective chloroplasts (Sato et al. 1995), and in a number of cyanobacterial species (Newton et al. 1979; Tyagi et al. 1992; Sinha & Hader, 2000). The effects observed include direct effects on membrane proteins and on a range of metabolic processes or indirect effects through the formation of reactive oxygen species such as superoxide (O^{2-}), hydroxyl radicals (-OH) or hydrogen peroxide (H_2O_2) (see review by Xue et al. 2005). Many microalgae, share with the larger algae, an ability to develop protective mechanism against the more damaging effects of UV radiation. Some cyanobacterial species, for example, possess active photoreactivation mechanisms which allow them to survive in the presence of intense solar UV-B radiation. Thus, Jiang & Qiu (2011) noted that short-term exposure of cultures of *Microcystis aeruginosa* to UV-B radiation (3.15 $W.m^{-2}$) severely inhibited photosynthetic capability, with the main damage centred on the oxygen-evolving complex of PSII. Return of the cultures to low-level photosynthetically active radiation (PAR) for a 20h period resulted in a rapid recovery (up to 90.9%). The recovery process, it was further shown, involved *de novo* protein synthesis..

The effects upon microalgae of increased exposure to UV radiation are inevitably linked to temperature effects, raising the possibility that these two effects may act either synergistically or antagonistically. This was explored by Halac et al. (2010) who, using the two diatoms *Chaetoceros gracilis* and *Thalassiosira weissflogii*, looked specifically for evidence for mitigation of the UV-induced photoinhibition of photosynthetic performance by exposure to higher temperature. They found that exposure to natural solar radiation at 18°C induced greater UV radiation photoinhibition in *T. weissflogii* than in *C. gracilis* but with both diatoms, exposure to a higher temperature (23°C) resulted in less photoinhibition. Heat dissipation processes were enhanced at the higher temperatures resulting in a shift in the balance between negative (photoinhibition) and positive (repair/dissipation) effects. In a mesocosm study of the combined effects of increased UV-B and temperature on natural phytoplankton assemblages, Lionard et al. (2012) noted that temperature plays a greater role than UV-B in structuring phytoplankton communities, possibly favouring diatoms over smaller species as the climate warms.

Among the most effective UV-protective mechanisms shown to exist in microalgae are those that involve development of screening pigments or energy dissipation mechanisms. Protection of cyanobacterial photosynthesis against UV-B damage has been attributed by Tyagi et al. (2003) to the role of ascorbic acid in maintaining the integrity of photosynthetic membranes. Another aspect of cyanobacterial metabolism shown to be severely affected by UV-B exposure is nitrogenase activity although here again, the effect was much reduced following the addition of ascorbic acid (Tyagi et al. 2003). Other UV-protective agents (effective in the cyanobacterium *Nostoc muscorum*) were L-cysteine, reduced glutathione, L-tryptophan and sodium pyruvate. Nitrogen limitation, on the other hand, has been reported to increase susceptibility of *Dunaliella tertiolecta* to UV-B damage (Shelley et al. 2002).

Many microalgal species, especially those occupying surface marine and freshwaters contain mycrosporine-like amino-acids (MAAs) and scytonemin whose formation is induced by exposure to UV radiation and for which a protective role is strongly suspected. Such compounds are effective in absorbing UV light, particularly UVB and UVA that occur at wavelengths of 280-320 nm and 320-400 nm, respectively, that would otherwise degrade organic molecules such as proteins and nucleic acids resulting in structural changes affecting biological function of the organisms concerned. MAAs are water-soluble compounds composed of a cyclohexane ring with an amino acid side group (Karentz et al. 1991)

The accumulation of MAAs is a widespread but not ubiquitous feature of marine phytoplankton species. Jeffrey et al. (1999) examined 152 species (206 strains) from 12 classes of microalgae for the presence of UVA- and UVB-absorbing compounds in cultures growing under visible light conditions (i.e in the absence of supplementary UV radiation). Measurement of the ratio of UV absorbance (280-390nm) to chlorophyll *a* absorbance (665 nm) allowed recognition of three groups of microalgal types: those with low UV:chl *a* absorbance ratios (0.18-0.9) (comprising diatoms, green algae, cyanobacteria, euglenophytes, eustigmatophytes, rhodophytes, some dinoflagellates and some prymnesiophytes), those in which this ratio has intermediate values (0.9-1.4) (comprising chrysophytes, some prasinophytes and some prymnesiophytes), and those with very high values for this ratio (1.4-6.75) (comprising surface bloom-forming dinoflagellates, cryptomonads, prymnesiophytes and raphidophytes). HPLC analysis of extracts of selected representatives of these groups showed the presence of a number of mycosporine-like amino acids, especially in those species with moderate to high UV:chl *a* absorbance ratios, with some of the bloom-forming dinflagellate species (e.g. *Gymnodinium catenatum*) showing results that would be consistent with survival under "sun-adapted" conditions.

There is evidence that induction of UV-absorbing compounds may be regulated by exposure to visible and UV light albeit with considerable variation between microalgal species in their capacity to respond. Riegger & Robinson (1997) examined the kinetics and spectral response of photoinductibility of UV-absorbing compounds in 11 Antarctic diatom species and in colonies of the prymnesiophyte *Phaeocystis antarctica.* High *in vivo* phytoplankton peaks (315-320 nm wavelengths), indicative of UV-absorbing compounds, were observed in all cultures of *P. antarctica.* Similar peaks at slightly longer wavelengths (330-333nm) were observed in some but not all the diatoms. The peaks were identified as the mycosporine-like amino acids porphyra-334, mycosporine-glycine and shinorine. In the diatoms, induction of these compounds was most effective at wavelengths of 370-460nm whereas in the prymnesiophyte, induction was maximal at 340nm and significant at wavelengths down to 305nm. The potential sunscreening effects were calculated as the fractional reduction of UV exposure and ranged from 0.03-0.50 for diatoms to 0.65-0.72 for *P. antarctica* colonies. It was concluded that the UV-absorbing compounds of *P. antarctica* may be located within the extracellular matrix where they are presumed to have a role in protecting this microalga from the elevated UVB radiation attributable to depletion of the ozone layer.

Klisch & Häder (2000) exposed cultures of the marine dinoflagellate *Gyrodinium dorsum* to polychromatic radiation (photosynthetically active radiation and UV) from a solar simulator for up to 72h, using cut-off filters to provide different irradiance spectra in the UV region. They found that exposure to UV irradiation of around 320nm induced an increase in the content of MAAs (normalised relative to chlorophyll *a* concentration) including several aminocyclohexenimine-MAAs and one aminocyclohexenone-MAA. In contrast, exposure to short-wavelength UV-B radiation caused a decrease in overall MAA production and a spectral shift in the absorption spectrum of the MAAs indicative of an altered MAA composition. They interpreted these observations as indicating the operation of an adaptive mechanism to the high irradiance in near-surface waters. The wavelength exposure most effective in stimulating MAA synthesis in *G. dorsum* was around 310nm (Klisch &

Häder,2002). PAR and long wavelength UV-A radiation showed almost no effect while wavelength below 300nm were severely damaging (as indicated by decreased chlorophyll *a* absorption).

The heterocystous nitrogen fixing cyanobacterium *Anabaena* sp.has only one type of MAA, shinorine, a bisubstituted MAA containing both glycine and serine groups and an absorption maximum at 334nm (Vincent & Roy,1993). Induction of the MAA, it was further shown, takes place only in cultures receiving irradiation in the UV range, peaking in the UV-B range at around 290nm. In another heterocystous cyanobacterium *Nostoc commune* the MAAs are located extracellularly and linked to oligosaccharides in the sheath (Ehling-Schultz et al. 1997).

Current, and anticipated likely future increases in UV radiation reaching the Earth's surface has raised concerns over possible deleterious effects on a range of ecosystems including effects on human health. The demonstrated role of MAAs in the natural defences of marine organisms (including microalgae) against UV radiation has suggested their possible wider use in other systems as protectants against UV damage. Current and anticipated climate changes resulting in thinner snow and reduced ice cover in polar regions might be expected to affect penetration of UV radiation with cosequent effects on sea ice microbial populations. Studies carried out at a Baltic Sea ice location in the Gulf of Finland showed that whilst the effects of UVA on bacteria was restricted to the top 5cm of ice, the effects on microalgae extended throught the ice column (Piiparinen & Kuosa, 2011). Extrapolating this observation to possible future climate changes would predict a decrease in ice-microalgal photosynthesis with flow-on effects on the food web. Effects of UVA on microalgal species composition, which were also noted, might be expected to influence nutrient and energy fluxes and the role of sea ice in seeding spring blooms.

Chapter III

Microalgae and Chemical Cycling in Marine and Freshwater Bodies

The microalgal species making up the phytoplankton communities of marine and freshwater bodies have the potential to influence various aspects of the physical and chemical characteristics of such bodies. This they do largely through their role as the primary producers at the base of the food chains and, in the case of some cyanobacterial species, through their ability to fix N_2. Research over the last few decades, however, has brought to light other effects attributable to some marine and freshwater microalgae that appear to influence strongly various biogeochemical processes in such water bodies and at their interface with the atmosphere.

The Carbon Cycle in Aquatic Ecosystems

Microalgae play a key role in the cycling of carbon and other inorganic nutrients in aquatic ecosystems. In marine waters, it has been estimated that approximately one half of the primary productivity can be attributed to diatoms, (Falkowski et al. 2000). Other major contributors are the picoplankters (0.2 to 2.0 μm cell size) (Johnson & Sieburth, 1979, Chisholm et al. 1988) and a wide range of eukaryotic microalgae including dinoflagellates, prasinophytes and prymnesiophytes (haptophytes). The picoplankters dominate the phytoplankton of the world's pelagic open oceans where their high surface area to volume ratio would be advantageous under the prevailing low nutrient conditions. In Red Sea waters, for example, the picoplanktonic cyanobacterium *Synechococcus* and the prochlorophyte *Prochlorococcus* together represent 69% of planktonic chlorophyll *a* and 77% of planktonic primary production (Lindell & Post,1995). Similarly high picophytoplankton contributions to primary production have also been reported for some freshwater ecosystems. They may, in some locations, reach abundances of 10^{6}-10^{8} cells ml^{-1} and contribute as much as 90-100% of the total microalgal biomass.

The microalgal populations (including the picophytoplankton) of oceanic waters are removed by the grazing of microzooplankton such as nanoflagellates and ciliates (Christaki et al. 2001) and by viral infection (Suttle & Chan, 1994). Similarly, in fresh-water ecosystems autotrophic picoplankton, whether they be prokaryotes or eukaryotes, are consumed by a range of heterotrophic nanoflagellates such as *Choricystis* sp., *Rhynchomonas* sp. and *Bodo saltans*. Grazing by flagellates has been shown to reduce significantly the stock of autotrophic picoplankton while, under appropriate conditions, also imposing a positive feedback on picoalgal growth (Bręk-Laitenen & Ojala, 2011).

Despite these grazing pressures, and despite their small size, a proportion of the picophytoplankton have been shown to sink to the lower meso- and bathypelic zones, perhaps following aggregation to form larger particles (Vanucci et al. 2001). Samples taken from depths of 200m and below in Suruga Bay (central Japan) and in the subtropical NW Pacific have shown abundances of the picophytoplankter *Synechococcus* proportional to the surface abundance (Sohrin et al. 2011), suggesting effective transport of these small cells down the water column with consequent major contributions to biogeochemical cycles in the dark ocean

Dissolved organic matter (DOM) in marine waters may be regarded as largely a byproduct of microalgal primary production (see Figure 3). It has been estimated that the DOM of marine waters is the largest pool of reduced carbon on the Earth's surface (Hedges, 2002). Transfer of organic matter from microalgae to the surrounding medium is the result of direct exudation by phytoplankton (Mague et al. 1980), or via heterotrophs (Nagata & Kirchman, 1992). It may also occur as a result of "sloppy feeding" (Lampert,1978) and through cell lysis or dissolution of faecal pellets and the formation of marine snow (Alldredge et al. 1993).

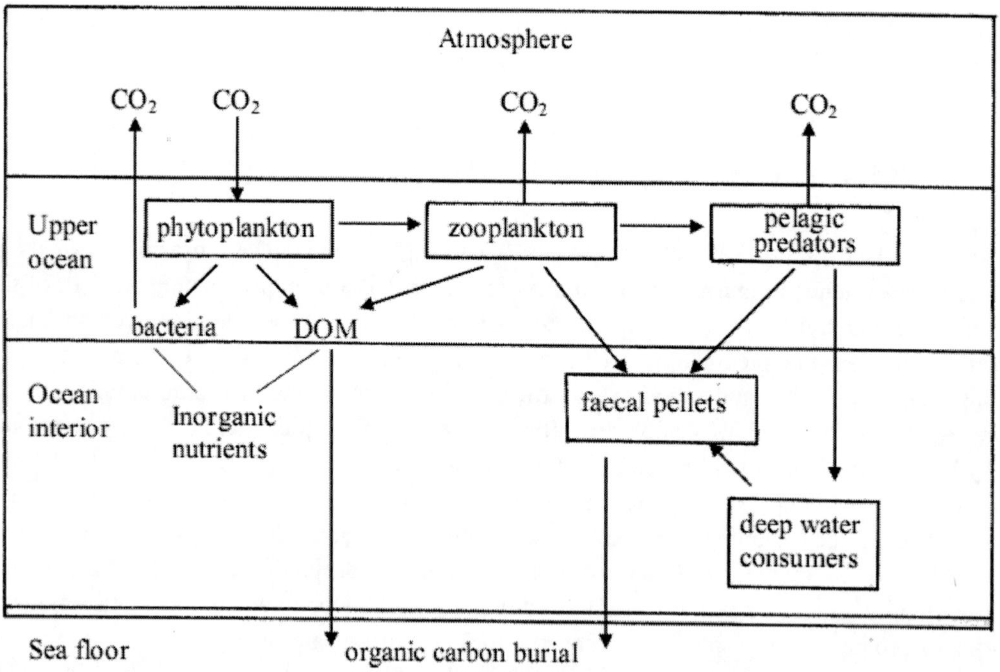

Figure 3. The cycling of carbon in oceanic waters from the phytoplankton primary producers to zooplankton grazers, bacteria and other pelagic heterotrophs, where most of it is consumed and respired, and ending up as dissolved organic matter thus enriching the inorganic pool below the thermocline and, through oxidation, maintaining the concentration of inorganic carbon in the ocean interior (Redrawn, with permission, from Falkowski & Oliver (2007), based on the original diagram of John Delaney, University of Washington, USA).

A significant proportion of the microalgal biomass consumed by copepods and various other mesozooplanton groups is released as fecal pellets. Some of this material will sink to the sea floor, but most will be degraded further by a number of heterotrophic or mixotrophic species, especially the larger heterotrophic dinoflagellates (Poulsen et al. 2011).

Aluwihare & Repeta (1999) investigated the high molecular weight fraction of the DOM released by laboratory cultures of two prymnesiophytes and a diatom and showed it to be rich in acyl heteropolysaccharides (APS) which were chemically similar to APS isolated from surface sea water and somewhat more resistant than other polysaccharides to degradation. Another phytoplankton mechanism that contributes significantly to the cycling of carbon in aquatic ecosystems is that which occurs when exopolymeric substances (EPS) are released from the cells. It has been estimated that up to 80-90% of the total extracellular release is in the form of polysaccharide material (Myklestad, 1995) that has the potential to play an important role in regulating sedimentation processes and biogeochemical cycling, by-passing the microbial loop. EPS has been shown to assist in the formation of aggregates when phytoplankton blooms decline (Logan et al. 1995) thus strongly affecting particle dynamics of a crucial phase of water column carbon cycling. Laboratory studies with the marine diatom *Amphora* have shown that EPS from that microalga can occur as particulate material (mostly of cell wall origin) and soluble material (i.e. exopolysaccharides), with the latter exhibiting strong binding characteristics (Zhang et al. 2008). A range of other compounds released by microphytoplankton can combine to form particulate matter rich in acidic polysaccharides, especially sulfated compounds which are extremely surface- active (Mopper et al. 1995). *Amphora*, for example, produces extracellular polymers capable of binding thorium (Zhang et al. 2008). A range of surface-active substances, largely derived from diatoms, include proteins, polysaccharides and lipids having both hydrophobic and hydrophilic functional groups and are hence able to participate in electrostatic and hydrophobic interactions.

Flux of these various forms of dissolved and particulate organic matter to the ocean interior and its subsequent oxidation enriches the inorganic nutrient pool below the thermocline. It maintains a high inorganic carbon concentration in the ocean interior, with a CO_2 content approximately 300ppm higher (by volume) than that of the atmosphere (Falkowski & Oliver, 2007). As a result of the various processes outlined above, only a very small proportion of the organic matter that sinks to the ocean interior finally becomes incorporated into the marine sediments to eventually form the sedimentary rocks.

Although most microalgae are photoautotrophs, many can adopt a mixotrophic form of nutrition in which photosynthetic CO_2 fixation is supplemented by the uptake and assimilation of carbon from a range of organic sources. Some others are entirely heterotrophic and can survive and grow in the complete absence of light, deriving both carbon and energy from organic substrates such as glucose, acetate or glycerol. Phagotrophy, that is, the ingestion of particulate matter ranging in size from bacterial size to particles (or microorganisms) 5μm in diameter, also occurs in some microalgae, especially some species of the haptophyte *Chrysochromulina* (Jones et al. 1994). Another group of microalgae with a special role in the cycling of carbon in marine waters are the zooxanthellae (mostly dinoflagellate species) which occur as symbionts within the tissues of a range of marine invertebrates including corals, clams, and zooanthids where the products of microalgal photosynthesis contribute to the carbon and other requirement of the host tissue. In scleractinian corals, for example, it has been estimated that autotrophic C acquisition (by the microalgal symbionts) covers more than 90% of the respiratory needs of the host (Tremblay et al. 2012). The importance of the microalgal symbionts to the coral, and to the reef as a whole, is reflected in the extreme damage resulting from so-called bleaching events caused by environmental stresses such as elevated seawater temperatures and irradiance (Hoegh-Guldberg, 2011).

For some microalgae, especially those in the picophytoplankton size range, the products of their photosynthesis enter the food chain through entrapment of the cells in mucus aggregates released by other biota. This applies particularly in reef ecosystems where the production of such aggregates allows removal of picophytoplankton biomass and its incorporation into taxa that do not possess the anatomical or physiological adaptations for the bulk capture of picoplankton (Naumann et al. 2009). Capture of large picoplankton-enriched mucus aggregates provides an important trophic source for a range of passive suspension-feeding organisms.

Nutrient Cycling in Aquatic Ecosystems

The cycling of nutrients between microalgae and their marine or freshwater environments is a function of their metabolic activities. In addition to the exchanges mentioned above relating to the fixation of carbon during photosynthesis, there are also exchanges involving other key nutrients. The so-called macronutrients nitrogen and phosphorus are vital components of proteins, nucleic acids and a number of other cellular metabolites such as those involved in energy exchanges. There are other nutrient requirements specific to certain microalgal species such as the silica requirement of diatoms, an indispensable constituent of their cell walls.

The source of nitrogen most readily available to microalgae is ammonium but most can also use nitrate via a nitrate reductase mechanism which involves reduction of the nitrate to ammonium. Nitrate is the major nitrogen source that becomes available to microalgae from land run off; ammonium usually represents a recycled form of nitrogen and becomes available to surface waters from upwelling of nutrient-rich deeper waters. Within the microalgal/phytoplankton component of aquatic ecosystems, the cyanobacteria are unique in their ability to fix nitrogen, i.e the conversion of gaseous nitrogen (N_2) to nitrogen in its reduced form ($-NH_2$) in a range of cellular organic compounds. The initial fixation process, catalysed by the enzyme nitrogenase, is of vital importance in all aquatic bodies because it can be a major source of new, usable nitrogen. A number of cyanobacterial genera have the ability to fix nitrogen. The most-studied of these so-called, diazotrophes, are the heterocystous forms where the nitrogenase is located in the heterocysts, specialised cells generally comprising 5-10% of the cyanobacterial biomass. In the heterocyst, which lacks the oxygen-yielding photosystem II component of photosynthesis, the nitrogenase is protected from the damaging effects of oxygen. The heterocysts also have a high capacity for respiratory O_2 consumption which, aided by their thick laminated cell walls, ensures a virtually anoxic interior environment.

The global distribution of the cyanobacteria generally and their ability to flourish in a variety of different habitats is well illustrated by the genus *Nostoc*. It is found in a variety of aquatic habitats and in terrestrial soils from the tropics to polar regions (Dodds et al. 1995). As a heterocystous, nitrogen-fixing genus, *Nostoc* plays a vital role in maintaining soil fertility in a vareity of terrestrial ecosystems including agricultural cropping systems. It often occurs in symbiotic associations such as in the lichens where, as the cyanobacterial component of the association, it provides both the photosynthetic primary production and nitrogen fixation.

Many, but not all, non-heterocystous cyanobacteria can also fix N_2, most of them only under strictly anaerobic conditions with only a few strains able to fix N_2 aerobically. In tropical and subtropical marine waters for example, the filamentous, non-heterocystous cyanobacterium *Trichodesmium* has long been recognised as a major contributor to pelagic nitrogen fixation (Capone et al. 1997, 2005), responsible for approximately 70% of annual nitrogen fixation in marine waters. It appears to do so by producing a small proportion of unusual cells that act as heterocyst equivalents even though they lack the structural features that in heterocysts help to protect against oxygen inactivation. A number of other non-hetrocystous cyanobacteria have been reported to share the ability of *Trichodesmium* to protect the nitrogen fixing enzyme from inactivation by O_2 although, unlike *Trichdesmium,* they do not fix N_2 during the day (Capone et al. 1990). The mechanisms involved in the integration of N_2 fixation in these cyanobacteria with other aspects of cellular metabolism (especially photosynthesis) has been reviewed by Bergman et al. (1997). Nitrogen fixation by *Trichodesmium* is often accompanied by low but significant release of hydrogen (see Chapter IX), with possible effects on atmospheric heat budgets through the formation of water vapour (Ehhalt & Prather, 2001). Nitrogen-fixing cultures of *T. erythraeum* were shown by Punshon & Moore (2008) to produce hydrogen at rates ranging from 0.06 to 0.71 nmol H_2 .μg chl.$a^{-1}h^{-1}$, suggesting that in oceanic waters this cyanobacterium may be a net source of hydrogen with consequent effects on the near surface atmosphere.

The occurrence of cyanobacteria as symbionts within a range of host organisms is clearly important in the nitrogen metabolism of the host cells and of the aquatic ecosystem to which they belong. Intracellular cyanobacterial symbionts occur in diatoms such as *Rhizosolenia,* which contains the non-heterocystous *Richelia intracellulari,* known to be a major nitrogen fixer (Rai, 1990). They also occur in heterotophic dinoflagellates (Farnelid et al. 2010) and, since they were isolated from oceanic waters (off Western Australia) characterised as having low nitrogen concentrations, the consortium (comprising eukaryotic microalga, cyanobacteria and heterotrophic bacteria) is presumed to have the capacity to fix gaseous nitrogen.

Molecular studies targeting the nitrogenase (*nifH*) gene encoding a component of the nitrogenase enzyme have shown that the ability to fix nitrogen is widespread among free-living unicellular cyanobacteria (e.g. Zehr et al. 2001). Similar *nifH* gene sequences have been detected in 23 of 37 isolates of the dinoflagellates *Ornithocerous* spp. and *Amphisolenia.* Interestingly, the overwhelming majority of the specimens contained *nifH* genes related to heterotrophic bacteria and the host dinoflagellate cells often contained a mixed assemblage of symbionts comprising non-diazotrophic cyanobacteria and N_2-fixing heterotrophic bacteria.

Microalgae play a key role in the cycling of phosphorus in aquatic ecosystems. Their incorporation into the food chain and their conversion into faecal pellets, delays the leaching of phosphorus that would otherwise occur from senescent phytoplankton cells (Tamelander et al. 2012). The smaller surface to volume ratio of aggregates and pellets compared with single cells and microalgal colonies means that less substrate is exposed for solubilization by alkaline phosphatases. Sinking of large zooplankton faecal pellets thus removes phosphorus from the water column as it does carbon and nitrogen. Such effects are likely to be particularly important under strongly stratified conditions when smaller, slower-sinking detritus material would be recycled (see Figure 3).

Trace Metal Cycling in the Ocean

Trace metals in oceanic waters exist predominantly in the form of metal ligands in which the metal (Cu, Cd or Fe) is complexed with a range of different organic molecules, a large proportion of which are microalgal in origin. Dissolved organic matter (DOM) plays several roles in the cycling of metals in aquatic environments. It can form stable metal/organic complexes thus reducing the toxicity of the free metal ions and it can act as an electron donor in the photoreduction and solubilisation of iron and manganese (Sunda, 1995). The amino acid component of DOM, in particular, has a high capacity for complexing free metal ions (Granum et al. 2002). Most amino acids can act as bidentate metal-binding ligands through their α-amino and α-carboxylate groups, with their side chains usually not involved in the metal binding process (Lorenzo et al. 2007). However, histidine and cysteine contain strong donor groups, imidazole and thiol groups in their side chains that can also contribute to complexation.

The copper-binding capacity of DOM produced by blooms of the marine diatom *Skeletonema costatum* was characterised by Lorenzo et al (2007) as comprising two types of ligands. One was related to labile DOC directly exuded by the microalga, the other to refractory humic-like material produced *in situ* by bacterial degradation of phytogenic material, with those two mechanisms operative at different stages of the bloom. Strmečki et al. (2010) characterised the nature and metal-complexing capacity of exudates from three marine phytoplankton species, the diatoms *Thalassiosira weissflogii* and *Skeletonema costatum* and the coccolithophore *Emiliania huxleyi*. The exudates were broadly classified as transparent exopolymer particles (TEP), surface-active substances (SAS) (derived largely from diatom exudates), reduced sulphur species (RSS) and thio/amino groups. Their complexing capacities for copper, cadmium and iron were investigated and compared with those of four selected marine polysaccharides (phytagel, carrageenan, laminarin and alginic acid). Exudates from the three phytoplankton species were particularly effective in complexing Cu. Comparisons with the complexing capacity of the tested polysaccharides (e.g. phytagel) indicated that the Cu was probably preferentially trapped within pores and channels of large hydrogels. Cd, a larger ion, was thought to be predominantly bound to the outer surfaces of gel particles. Thio/amino groups of sulphur-rich 'glutathione' type ligands were found in all phytoplankton cultures and, of all the substance classes tested, were deemed to be the most effective in complexing Cu.

Volatile Microalgal Products

A number of marine microalgal species, have been shown to produce the sulphurous gas dimethylsulfide (DMS) $(CH_3)_2S$, which escapes from surface oceanic waters to the atmosphere (Malin & Kirst 1997). It has been estimated that about 50% of the biogenic sulphur reaching the atmosphere does so as DMS (Malin & Grimalt 2000). The most prolific microalgal producers of this volatile gas are certain species of dinoflagellates and prymnesiophytes that accumulate the precursor compound dimethylsulphoniopropionate (DMSP) within their cells. The physiological role of DMSP is primarily as a neutral solute that regulates an osmotic balance between the microalga and the surrounding medium, although it may also function as an antioxidant (Sunda et al. 2002) or in chemical defence against grazing (Wolfe 2000). Production of DMS is thought to be the result of damage of the microalgal cells through the action of zooplankton grazing or physical stress or turbulence and it may also involve bacterial metabolism of the released DMSP. In cultures of *Emiliania huxley*, there was a direct link between photo-oxidative stress (high photon flux densities plus UV radiation) and net production of dissolved DMSP and DMS (Archer et al. 2010).

DMSP also undergoes enzymatic cleavage, catalysed by the enzyme DMSP-lyase which has been reported for a number of bacterial isolates (de Souza & Yoch 1995). This is thought to be the dominant mechanism for DMS production in some marine environments (Kiene & Linn 2000). A number of microalgal species also possess DMSP-lyase activity. Niki et al. (2000) found DMSP-lyase activity in strains of *Heterocapsa triquetra* and *Scrippsiella trochoidea,* both DMSP-producing dinoflagellates, although two other DMSP-producing microalgae, the prymnesiophytes *Isochrysis galbana* and *Gephyrocapsa oceanica* did not display any measurable DMSP-lyase activity. Neither, unsurpisingly, was there any measurable DMSP-lyase activity in the non-DMSP-producer *Heterostigma akashiwo* (Raphidophyceae).

The released DMS may oxidise to form sulphate aerosol particles that act as nuclei that attract molecules of water to produce clouds. These may have the effect of scattering incoming solar radiation or blocking outgoing infrared fluxes with consequent effects on sea-surface temperature and hence on climate (Charlson et al. 1987). Gondwe et al. (2003) have estimated that DMS may be responsible for 43% of climate-relevant, sea-salt sulphate in the relatively oligotrophic southern oceans and model simulations have pointed to strong feed-back effects of cloud droplet formation involving marine biota.

Of the microalgal groups known to have high intracellular concentrations of DMSP, and hence a high potential for high DMS release to the atmosphere, the coccolithophores (Haptophyta) have received particular attention. Franklin et al. (2010) measured the cellular concentration of DMSP and DMS in ten cultured coccolithophore species as well as their DMSP-lyase (DLA) activity. Intracellular concentrations of DMSP ranged from 261mM in *Gephyrocapsa oceanica* to 715mM in *Umbilicosphaera hulburtiana*. The mean DMSP quota was lowest in a strain of *Emiliania huxleyi* (0.8 pg.cell $^{-1}$) and highest in *Coccolithus braarudii* (95pg.cell $^{-1}$). There was a significant correlation, in all the species tested, between the cell quota of DMSP and cell volume, suggesting a fundamental physiological role for this compound as a compatible solute. The enzymic conversion of DMSP to DMS by DLA has the potential to influence DMS concentrations in oceanic waters although, of the coccolithophore species tested by Franklin et al. (2010), DLA was restricted to only two species namely *Emiliania huxleyi* and *Gephyrocapsa oceanica.* Clearly, marine ecosystems with a significant coccolithophore component are likely to have the capacity for substantial DMS production whether it be through direct release to the atmosphere or indirectly via a microbial loop.

The interactions within the marine microbial food web that lead to release of DMS to the atmosphere are summarised in Figure 4. It is suggested that although production of DMSP and DMS can occur by microalgal exudation and natural senescence, zooplankton grazing is also likely to be a major contributor. Viral-induced lysis of microalgal cells has also been implicated in DMSP release and in its rapid conversion to DMS (Malin *et al.*1998).

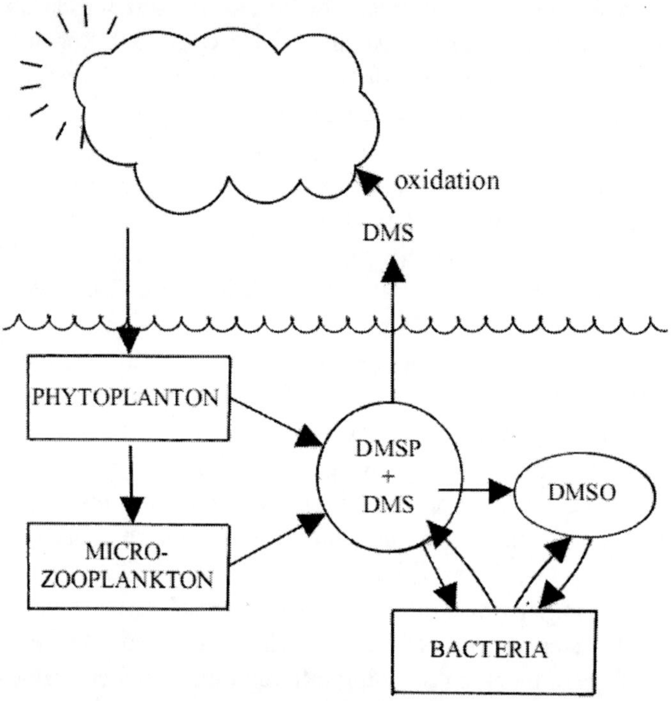

Figure 4. Scheme depicting the production of dissolved DMSP and DMS directly from microalgae or indirectly via the marine microbial food web, and the release of DMS and DMSO to the atmosphere so affecting climate and atmosphere-upper ocean exchanges. (Redrawn, with permission, from Malin et al., 1994).

DMS may also be subject to abiotic photooxidation leading to a build up of dimethyl sulphoxide (DMSO) in the surface waters or in the atmosphere immediately above (Brimblecombe & Shooter 1986). Turnover of the various methylated sulphur compounds was found by Simo et al. (2000) to vary depending on the phytoplankton standing stock. DMS production and DMSP consumption were particularly high during a *Phaeocystis* sp. bloom in a North Sea site. There was strong evidence for biological cycling of DMSP with between 9 and 96% of the DMSP consumed being converted to DMS. Similarly, in subpolar regions of the North Atlantic, DMSP and its derivatives accounted for between 48 and 100% of the sulphur fluxes and between 5 and 15% of the carbon fluxes, with the microzooplankton and heterotrophic bacteria playing a major role in controlling the amount of phytoplankton DMSP that is released to the atmosphere as DMS (Simó et al. 2002).

DMS has been described as a "keystone" molecule – a chemical that has a disproportionate effect on the environment relative to its abundance. It has been suggested. That DMS released from microalgae (e.g. during crustacean feeding) can be sensed by predators higher up the food chain whose excretory products in turn nourish further microalgal growth so sustaining nutrient cycling in marine ecosystems.

Among the potentially harmful products produced by microalgae are certain volatile halogenated organic compounds (VHOC) in which some of the carbon atoms of the molecule are linked by co-valent bonds with fluorine, chlorine, bromine or iodine atoms to form, respectively organofluoride, organochloride, organobromide or organoiodide compounds. Marine microalgae, in particular, produce several methane and ethane derivatives with the potential to form marine aerosols and to degrade ozone in the stratosphere. Production rates for $CHBr_3$ by microalgae in southern oceanic waters has been shown to vary between 6 and 6000 pmol [μg chl *a*]$^{-1}$ day^{-1} (Abrahamsson et al. 2004). A range of diatom species have been shown by Moore et al. (1996) to produce halocarbons at rates generally higher than those obtained by other microalgae. The increased occurrence and intensity of cyanobacterial blooms in the Baltic Sea in recent decades has been accompanied by increased production of volatile halogenated organic compounds (VHOC) with the potential to degrade stratospheric ozone (Solomon et al. 1992). Incubation experiments with samples taken from a summer Baltic Sea cyanobacterial bloom (dominated by *Nodularia spumigena*) showed significant production of $CHBr_3$, CH_3I, CH_2Br_2 and $CHBr_2Cl$, with production rates up to 0.3 pmol [μg chl *a*]$^{-1}$h^{-1} of the first named halocarbon at mid day (Karlsson et al. 2008).

Bacterial and Viral Effects on Microalgae

An often overlooked component of the complex of interactions between microalgae and their aquatic environment is their susceptibility to viral infection, a susceptibility they share with bacteria and microzooplankton (Fuhrman, 1999; González & Suttle 1993). Microalgal viral infections have, for example, been implicated in the decline of phytoplankton blooms (Suttle, 2000) and are generally considered to be influential in the transfer of microalgal (and other) biomass into the pool of dissolved organic matter, so diverting organic matter flow away from higher trophic levels in what has been termed the 'viral shunt'. It has been suggested that viral-mediated mortality of host cells, by enhancing nutrient regeneration processes, has the potential to contribute to sustained phytoplankton growth in surface waters (Brussard et al. 1996).

A number of studies have indicated that viral lysis of host cells may release a range of labile cellular components including nutrients, nucleic acids and trace elements (see review by Danovaro et al. 2010). Interestingly, some of the highest recordings of viral infections among phytoplankton species have come from the cyanobacterium *Synechococcus,* a prominent contributor to the primary production of oceanic waters (Proctor & Fuhrman, 1990). High levels of mortality attributed to viral infection have also been recorded for the haptophyte *Emiliania huxleyi,* the diatom *Heterosigma akashiwo* and the chrysophytes *Aureococcus anophagefferens* and *Phaeocystis globosa.* Virus-mediated mortality of the picoeukaryotic phytoplankter *Micromonas pusilla* (Prasinophyceae) is reported to cause a turnover of the host standing stock of between 2 and 25% day^{-1}(Cottrell & Suttle, 1995) while the cyanobacterium *Synechococcus* is said to lose 1-8% of its population daily (Suttle & Chan 1994). Viral lysis of the marine chrysophyte *Aureococcus anophagefferens* has been shown to release iron which can be rapidly assimilated by other plankton and it has been estimated that virus-mediated iron release can support as much as 90% of the primary production in high-nutrient, low-chlorophyll oceanic regions such as those of subtropical equatorial eastern Pacific (Poorvin et al. 2004). Consequently, since much of the primary production of oceanic waters is, at least periodically, limited by iron availability (Moore et al. 2002), viruses clearly make an important contribution to oceanic carbon cycling.

In an interesting link with a feature of microalgal biology described above, it has been suggested that strains of *Emiliania huxleyi* with a high capacity for converting DMSP to acrylic acid and DMS (i.e. high DMSP-lyase activity) are particularly resistant to viral infection suggesting that the DMSP system in phytoplankton generally may have as one of its primary functions, a chemical defence against viral infection (Evans et al. 2006). Viral infection is also recognised as a significant contributor to the mortality of benthic microalgae. Hewson et al. (2001) showed that experimentally induced viral infection of sediment microalgae can result in a 90% decrease in the biomass of benthic pennate diatoms and a 20-26% decrease for euglenophytes. It is clear, therefore, that viral infections of microalgae (and other biota) have the potential to influence greatly the cycling of carbon and other nutrients in aquatic ecosystems with major effects on their carbon sequestration capacities (see Section XI below).

A number of microalgae have been shown to be susceptible to algicidal substances of bacterial origin, especially in eutrophic lakes where such interactions have been shown to be important in the decline of cyanobacterial blooms (Nakamura et al. 2003). The bacterium *Bacillus cereus* has been shown to induce high microalgal lysis of *Anabaena flos-aquae* under alkaline conditions, attributable to a novel algicide characterised as being hydrophilic, polar and thermotolerant (Zhang et al. 2010). Exposure of this cyanobacterium to low concentrations of the *B. cereus* algicide stimulated growth and a range of metabolic processes while at higher concentrations the effects were strongly inhibitory and shown to be associated with antioxidative/oxidative reactions (Zhao et al. 2012).

Chapter IV

Mass Culture of Microalgae

Any program for the mass culture of microalgae will have as its main objective the production of a defined biomass of microalgal material raised under conditions regulated to suit the particular requirements of the final product. The methods used can range from simple ponds or raceways to highly mechanised photobioreactor systems. When the biomass is to be used for the extraction of specific high-value compounds, the original biomass can be raised on a large scale in open ponds under close to natural conditions provided the level of contamination by other species does not seriously affect production or does not introduce undesirable compounds that would be coextracted with the product compound. Where greater control is required over the purity of the culture and over the conditions under which the microalgal cells are raised, the biomass has to be raised using methods more akin to fermentation technology. Only a relatively small number of microalgal species have thus far been successfully produced commercially on a large scale. They include the chlorophytes *Chlorella* and *Dunaliella* and the cyanobacterium *Spirulina* (*Arthrospira*) all of which, for different reasons, can be grown under conditions that minimise contamination by other microorganisms. Many diatoms and some dinoflagellates are also grown, usually on a smaller scale and usually as natural feed to support aquaculture, while some other species are grown because they produce some specific high-value compounds, a good example being *Haematococcus pluvialis*, a rich source of carotenoids. The rough biomass estimates published by Pulz & Gross (2004) place *Spirulina* and *Chlorella* at the head of quite a short list with global productivities, at the time of that study of 3,000 and 2,000 t/year respectively. The Aquatic Species Program of the U.S. Department of Energy carried out at Roswell, New Mexico over three decades ago reported maximum recovery rates of 50g microalgal dry matter $m^{-2}d^{-1}$ (Sheehan et al. 1998). More recent studies have estimated equivalent theoretical recovery rates of over 100g using appropriately illuminated photobioreactors (Gordon & Polle, 2007), well in excess of the practical maximum biomass production of $38g.m^{-2}.d^{-1}$ reported by Eriksen (2008).

Mass cultures are usually initiated from “seeding material” derived from pure microalgal cultures established by careful isolation and purification to provide unialgal cultures free of any bacteria or other contaminating organisms. The more traditional methods of manually picking and isolating individual cells of the desired species can, for some species, be replaced by less labour-intensive electronic sorting methods used in conjunction with flow cytometry. The latter methods have been successfully used for the isolation of several microalgal species from mixed culture assemblages and directly from natural samples (Sieracki et al. 2005). Sinigalliano et al. (2009) used multiparametric flow cytometry to sort several dinoflagellate species and achieved success rates comparable with the traditional manual methods. In particular, they devised a highly restrictive gating technique to select and separate cells of *Karenia brevis* from a natural bloom sample. Such methods, when combined with suitable improvement of post-sorting and maintenance methods have the potential, they conclude, to accelerate the discovery of new microalgal strains thus removing some of the current impediments to their wider use in biotechnological applications (Shimizu & Li 2006).

Microlagal mass cultures may be raised either as batch cultures or as continuous cultures. In the former, an aliquot of the "seed" culture is inoculated into a large amount of a suitable medium containing, in excess, all the requirements (including light) for growth of the microalga over a number of generations. After an initial period of adjustment, the microalgal cells will divide exponentially but eventually, growth will be limited by the availability of some key requirement (light, nutrient or other factor) and the culture will move into a stationary phase. In continuous cultures, the various requirements for growth are supplied continuously and cells removed at a constant rate equal to or less than the rate of production of new cells. Continuous cultures may be classified as chemostats, in which the rate of growth is controlled (at some level below the maximum) by the rate of supply of a limiting nutrient, or turbidostats where the rate of growth is governed by the rate of supply of fresh medium which in turn is governed by changes in turbidity of the culture. A review of the practical and economic feasibility of the commercial exploitation of microalgal mass culture technologies as they were known before the more rapid developments of recent decades is given by Soeder (1980).

Ponds and Raceways

The simplest form of large- scale production of microalgal biomass is the shallow open pond system with natural illumination with or without some form of mixing to increase gas exchange between the culture and the atmosphere. Physical mixing with foils or paddles is usually considered to be more effective than more vigorous aeration using a gas stream with or without enrichment with CO_2. A good example of a low cost open pond system is that used in Western Australia for the culture of *Dunaliella* for ß-carotene production in large, shallow ponds (Borowitzka et al. 1984; Borowitzka, 1991). The success of this system has been attributed to its simplicity (no artificial mixing, other than by wind or convection), its large scale (ponds of up to 250 ha in area and 50cm or less depth) and its use of locally-available sea water as the growth medium with high salinities, tolerable to the cultured species but not to other contaminating species. Other contributing factors include the suitability of the climate for year-round production and, most importantly, its location on land not otherwise suitable for agricultural/horticultural production. A refinement of the open pond system is the raceway system, which allows continuous flow of the culture, usually driven by paddle wheels. A further refinement is the cascade sytem, first developed by Setlik et al. (1970) which, by arranging for rapid flow of the culture (in some cases over a sloping surface and at depths as little as 1cm or less), can yeld high cell densities and good rates of growth (Doucha & Livansky, 1995).

Efficient large-scale pond-culturing of microalgae is possible only if a number of problems can be overcome. First, there is the problem of oversaturation of photosynthesis (the normal condition on a sunny day). A possible solution to this problem would be the development of new strains with smaller antenna sizes (Melis, 2009). Another would be to increase turbulence, with inevitable cost implications. Continuous production of large quantities of microalgal biomass has a high requirement for CO_2 It has been estimated that production of 1 ton of microalgal biomass requires 1.8 tons of CO_2 (Kliphuis et al. 2010), pointing to a possible contribution of mass microalgal photosynthesis towards alleviating increasing atmospheric CO_2 concentrations (see Chapter XI below). The requirement for essential nutrients (primarily sources of N and P) would also be substantial, and in competition with existing agricultural and horticultural needs. A large-scale microalgal mass culture sytem would be most feasible when linked with an efficient system of nutrient re-cycling. There would be similar constraints in relation to water usage.

Shimamatsu (2004) described some of the features of two open pond commercial facilities for the mass culture of the cyanobacterium *Spirulina*, one in Thailand (2,000m^2 in area) and a larger one in California, USA (5,000m^2) (see Figure 5).

Each pond is rectangular and fitted with a central partition running along most of the length of the pond but leaving end gaps to allow circulation of the culture, driven by a paddle-wheel system situated across one half of the pond. There are also a series of corner devices to aid circulation. Culture depth is maintained at 15-39cm and microalgal density at 400-600 mg. dry weight L^{-1}. Among the problems encountered were those related to contamination by other microrganisms, especially following dilution by rain. There were also problems relating to harvesting and drying of the product and ensuring that it conforms with the necessary quality standards.

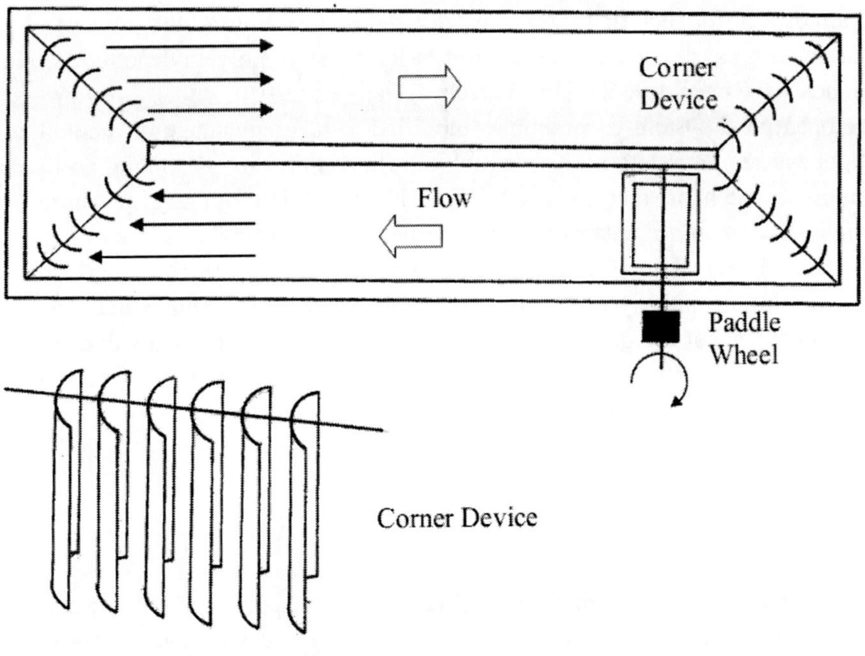

Figure 5. Design of a typical open pond system for microalgal mass culture showing a paddle wheel, central partition and corner devices to assist circular flow. (Redrawn, with permission, from Shimamatsu, 2004).

Solar salt works comprise a series (usually about ten) connected shallow ponds of increasing salinity (from sea water up to ca. 280 -285‰) before the flow-through is fed into the final crystallizers. They provide conditions allowing the development of a number of special microalgal populations selected by the prevailing conditions. The microalgal flora of the higher salinity ponds is usually dominated cyanobacteria where species such as *Aphanothece halophytica* and *Microcoleus chtonoplastes* form the basis of compact microbial mats (Alfinito et al. 1990). In the lower salinity ponds (up to 110‰) the microalgal flora is largely dominated by a number of euryhaline diatoms. The ability of these microalgal species (and some others) to survive under such extreme conditions raises interesting possibilities with regard to their development of protective mechanisms, particularly in relation to the production of compounds of potential commercial value (see Chapter VI). As far as the cyanobacteria are concern, their effect on salt production is more likely to be negative since the extensive mucilage production can adverseley affect the salt crystallisation process with consequent serious losses during the salt washing stages.

Mass cultures based on the use of plastic tubes share many of the advantages of the open ponds or raceways (short light path etc) while at the same time allowing for exclusion of contaminants and providing some control over environmental conditions. Scaled-up versions, using large plastic bags (of about 0.5m diameter, holding up to 1000 L) have been used either in batch mode or as semi-continuous systems, to provide microalgal biomass in the aquaculture industry (Watson, 1979). Comparison of the growth of the Haptophyte *Pavlova lutheri* and the diatom *Chaetoceros muelleri* in 250L bags showed that the former grew better in narrow bags (with the better light conditions) while the latter grew better in wider bags (under light limitation) (Jacobsen et al. 2012). The use of artificial illumination to supplement sunlight can counter the diurnal or seasonal unreliability of natural sunlight but is only really practical on a relatively small scale or if the product is of high value (such as in the production of feed for specific stages of hatcheries or other aquaculture projects). Some other low-cost culturing systems based on the use of long-lasting polyethylene tubes, immersed in water ponds, have been described by Wijffels & Barbosa (2010).

The open pond system is sometimes modified to take advantage of the ability of many microalgal species to undergo a heterotrophic form of nutrition by linking on to a source of organic matter such as that from agricultural waste. Heterotrophy, or more commonly mixotrophy, occurs in a number of green microalgal species and in some cyanobacteria and diatoms. Among the latter, *Phaeodactylum tricornutum,* has received particular attention because it lacks the heavily silicified cell wall characteristic of other diatoms. It is either unsilicified (except for the girdle bands) or has a single small silicified valve (Borowitzka & Volcani, 1978) which, together with its high content of polyunsaturated fatty acids (PUFAs), makes it suitable for a range of mass-culture applications.

Photobioreactors

Photobioreactors (PBRs) are usually closed units designed specifically to suit a particular microalgal species and to provide conditions that maximise the yield of the product to be harvested. They ensure reproducibility of the conditions and hence of the product while at the same time reducing the risk of contamination by other organisms. Closed PBRs would be essential for the culture of microalgae genetically modified to increase the yield of specific high-value compounds. They offer the best way of applying strict regulation to prevent any potential threats to natural ecosystems.

For autotrophic cultures, the major limiting factors will be the provision of adequate gas exchange and mixing of the culture. Of particular importance will be the need to ensure an adequate supply of light and carbon dioxide. The latter of these can be met by maintaining the pH and bicarbonate concentration at optimum levels. In a batch culture, the pH will tend to increase with growth of the culture and this, in turn, will reduce the availability of inorganic carbon. The usual procedure to counter this effect is to aerate with an air stream with enhanced CO_2 partial pressure. Another method is to use a concentrated carbonate buffer as a CO_2 source, usually in a separate chamber as in the Warburg instrument for measuring photosynthesis. Pörs et al. (2010) devised a system of providing CO_2 to photoautotrophic cultures of cyanobacteria and unicellular algae from small polyethylene bags acting as CO_2 reservoirs inserted within the culture vessel. The bags contained a concentrated carbonate buffer with a CO_2 partial pressure of 32 mbar which released the CO_2 through the membrane directly into the medium. It was found that the CO_2 reservoirs were not depleted over a period of one week and growth of a number of the microalgae tested was exponential over the first two days suggesting that such diffusive carbon supply methods may, with suitable modification, be suitable for biotechnological applications.

The problem of ensuring a suitable supply of light to the individual cells of the culture is of particular concern in the design of any autotrophic culture apparatus. It is this requirement, and in particular the exponential attenuation of light as it passes through a liquid culture, that primarily influences the design of the culture vessel. The conventional light curve relating growth (or photosynthetic rate) of a culture to light intensity at the culture surface, holds only for optically thin cultures, but microalgae cannot be exposed for long to such high light intensities without suffering extreme photoinhibition. Richmond (2004) has summarised the principles that have to be considered in any attempt to achieve maximal microalgal productivity in a photobioreactor. They include various methods for keeping the culture at the optimum cell density such as the imposition of light-dark cycles, efficient mixing of the culture (in cultures exposed to high light) or removing the inhibitory effects on cell growth (in ultra high-density cultures).

The way that these requirements have been addressed in different photobioreactor design types has been decribed by Tsygankov (2001). Some of the key features of each of these design types are summarised here:

Cuvettes, Tubes, and Cylinders

Transparent cuvettes provided with an external source of illumination (from one or both sides) also require a CO_2 supply, usually bubbled from below as a gas mixture. The optimum thickness of the cuvette is largely a compromise between growth suppression in the deeper layers (due to insufficient illumination) and in the surface layers (due to the inhibitory effect of excess illumination).

Among the drawbacks of this system are problems relating to inadequate mixing, settling within the culture and the formation of surface films. The maintenance of sterile conditions is also difficult.

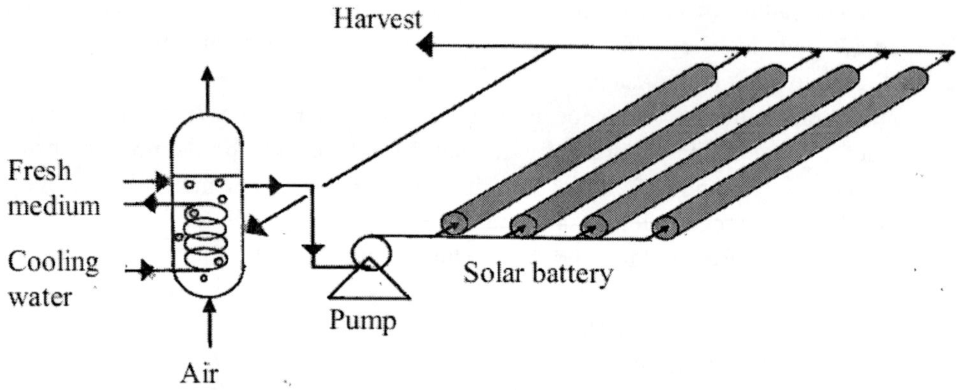

Figure 6. Design of a typical tubular photobioreactor for microalgal mass culture. Fresh medium adjusted for pH, temperature, nutrient and gaseous supply is pumped from the main vessel through a bank of illuminated parallel plastic or glass tubes and returned to the main vessel or diverted for harvesting and processing, (Redrawn, with permission, from Varfolomeev & Wasserman, 2011).

In the tube- or loop-reactors, glass, PVC or fluoroplast tubes containing the culture are connected in series to a gas mass-exchanger and illuminated by an external light source which may be positioned within the cylinder formed by the coiled tubes. Such systems often involve spatial separation of photosynthesis (in the tubes) from the gas exchange component. Beacause if this, each cell of the cultured organism is exposed to fluctuations of a range of cultural conditions. There are also a number of other drawbacks but the tube PBR does allow some upscaling through the inclusion of additional tube units. A typical example of a tubular photobioreactor is that shown in Figure 6. It consists of a bank of parallel plastic or glass tubes (up to tens of metres in length), through which the microalgal culture is pumped after passing through a supply vessel with facilities for controlling pH, temperature and gaseous concentrations. Photobioreactors of this type, working under optimal conditions (and with artificial light) have been reported to support productivity of 120-150g microalgal biomass.m^{-2} day^{-1}(Tsoglin & Gabel, 2000). Chisti (2007) has presented data comparing the yield of microalgal biomass obtained from an open cultivation system with that from a closed phytobioreactor. The main difference between the two culture systems is the higher biomass concentration (and higher lipid yield) obtained with the photobioreactor and a lower requirement of liquid medium compared with a corresponding open cultivator. Some systems use two coaxial transparent cylinders providing space for the culture between them and illumination by an external light source or one mounted within the common axis of the cylinders. The system is compact, amenable to mechanical mixing, easily sterilised but does not lend itself to easy scaling up.

Fermentor-Type Reactors

Closed photobioreactors are particularly suited for the culture of microalgal species capable of adopting a heterotrophic or mixotrophic form of nutrition using acetate, glucose or some other organic sustrate. Heterotrophic cultures of *Chlorella* have been grown in Japan with high yields (Lee, 1997) and the dinoflagellate *Crypthecodinium cohnii* has been grown heterotrophically for the production of long-chain polyunsaturated fatty acids (Kyle et al.

1998). To cater for the special needs of photosynthetic organisms, however, industrial scale bioreactors designed for the culture of non-photosynthetic organisms require extensive modification. Detailed accounts of some of the criteria used in the design of PBRs are provided in reviews by Pulz (2001), Tsygankov (2001), Tredici (2004) and Carvalho et al. (2006). For a variety of reasons, none of the designs so far attempted have been completely successful and they all have relatively small working volumes. Some of the modifications that have been attempted include the use of solar energy concentrators, the use of more efficient light sources such as light-emitting diodes (LED) (Görs et al. 2009) and the organisation of the illuminated part of the reactor as a flowing film.

Fermenters of 50, 200 and 600L were used by Doucha & Livanský (2012) for growing heterotrophic cultures of *Chlorella vulgaris* in aerated and mixed nutrient media supplemented with urea (as a nitrogen source) and glucose (as a source of carbon and energy). They achieved culture densities above 100g DW L^{-1}, and growth rates of 3.52g $DWL^{-1}h^{-1}$ (in a 50 L fermenter) and 3.37 DW $L^{-1}h^{-1}$ (in a 200L fermenter). In the 600L fermenter, the growth rate was only 1.22 gDW $L^{-1}h^{-1}$ due, it was shown, to the failure to maintain adequate dissolved oxygen concentration. The maximum specific growth rate (μ) attained was ca. 0.15 h^{-1}, about 10 to 20 times higher than those previously reported for phototrophically grown *Chlorella* cultures. Significantly, allowing the cultures to proceed beyond the point of glucose sufficiency increased the cellular content of proteins, ß-carotene and chlorophylls and decreased the starch content.

Photobioreactors, whatever their design, are most suited to small-scale, specialist microalgal culture with limited scope for upscaling to mass culture. Specialist needs are best met by tailoring the design of the culture system to suit the microalgal species, the product to be harvested and the prevailing conditions. This is well illustrated by the design proposed by Masojídek et al. (2009) which comprises a novel two-panel solar photobioreactor system which allows maximum flexibility of solar irradiance for a temperate climate zone (Centre for Biotechnology, Nové Hrady, Czech Republic). The two panels of each unit occupy, respectively, the vertical and inclined south-facing walls of a greenhouse-like construction. Each panel consists of glass tubular cultivation loops through which the microalgal culture circulates under conditions regulated by appropriate control mechanisms and associated software. Supra-high solar radiation is provided by a system of lenses, which concentrate the radiation on to the surface of the glass cultivation tubes. The vertical collector panels were shown to be most effective in spring and autumn when the sun inclination angle is low; while the inclined panels received maximum irradiance in summer.

Using the cyanobacterium *Arthrospira platensis* as a test organism, Masojídek et al. (2009) achieved high growth rates even at irradiance levels higher than those usually recorded outdoors in summer. The system was also used to investigate the production of the carotenoid astaxanthin (see Section VI, below) by the green microalg *Haematococcus pluvialis*; in particular, to explore the correlation between photochemical activities and production of the carotenoid. The culture was first grown in low-irradiance units, and then exposed to supra-high irradiance which stimulated increased astaxanthin production, reaching 3% of dry weight within four days.

In any microalgal culture, exposure to low light intensity allows the antenna capacity to function close to its maximum, and almost all of the absorbed energy is funnelled to the reaction centre to drive the photochemical reactions. At high irradiance, certain components of the antennae act as energy dissipators. In eukaryotic microalgae, this process involves the membrane-bound chlorophyll-containing light –harvesting complexes of PSII and is triggered by acidification of the thylakoid lumen under saturating light conditions (Horton et al. 1996). In cyanobacteria, the process is induced by the photoactivation of a soluble carotenoid protein, the Orange Carotenoid Protein (OCP) and involves soluble extramembrane antennae complexes, the phycobilisomes (PBs) which carry blue and red phycobiliproteins. Work with *Synechocystis* (Gwizdala et al. 2011) has shown that one OCP per PB is sufficient to quench almost all PB fluorescence, indicating that it is a very powerful mechanism for decreasing the energy arriving at the reaction centres. Cells of this cyanobacterium grown under low or medium light conditions quench only 30% of the maximal fluorescence and only 30% less energy arrives at the photosystem. Under stress conditions, such as iron starvation, the OCP to PB ratio increases markedly and 60 to 70% of maximal fluorescence is quenched (Wilson et al., 2007).

A number of microalgal species have been grown successfully in culture systems where the individual cells are held on to a solid matrix over which flow of the culture fluid can be closely controlled. Cells of *Chlamydomonas reinhardtii*, immobilized by entrapment within beads of calcium alginate, have been shown to retain their photosynthetic activity and to grow with generation times comparable with those of free cells under similar conditions (Santos-Rosa et al. 1989). Alginates, composed of a mixture of mannuronic and glucuronic acids extracted from certain marine macroalgae, provide a three dimensional structural matrix which is biochemically inactive and whose interstic spaces can be varied depending on the type of alginate and the metal ion (Ca^{2+}, Ba^{2+}, Sr^{2+}) used for making the gel (Galván & de la Rosa, 1997). Bioproduction of ammonium, glycerol and glycolate by alginate-immobilised *C. reinhardtii* has been shown by Galván & de la Rosa (1997) to duplicate the production levels obtained with free-cell cultures. Immobilization systems have also been used for some cyanobacterial species such as *Anabaena azollae* held in synhetic polyvinyl polymer foam fed by trickling culture medium.

Harvesting Methods

For the commercial production of microalgal biomass, the cost of harvesting, and processing (including de-watering) is one of the major impediments to economically viable production of bulk food, feed or biofuel. It has been estimated that recovery processes may account for at least 20-30% of the total cost of producing microalgal biomass (Gudin & Thepenier, 1986). Some of the problems relate to the small size of most microalgal species and the small difference in density between the microalgal cell and its medium. Efficient harvesting therefore requires centrifugation, filtration or flocculation. An evaluation of several harvesting methods has shown that flocculation combined with floatation or sedimentation and subsequent further dewatering by centrifugation or filtration are the most cost- and energy-efficient procedures (Schenk et al. 2008).

Numerous chemical coagulants or flocculants have been tested including metal salts (aluminium sulphate, ferric chloride, ferric sulphate etc) and cationic polysaccharides such as chitosan (Zeng et al. 2008). Some microalgal cultures can be induced to flocculate simply by stopping aeration (auto-flocculation). This is due to precipitation at elevated pHs of carbonate salts along with the microalgal cells (Sukenik & Shelef, 1984). Microalgal cultures generally experience increases of pH during exponential growth due to photosynthetic CO_2 consumption and such auto-flocculation effects can also be induced by the addition of NaOH (Chen et al. 2011). Sirin et al. (2011) explored the efficiency of various methods for flocculating cultures of *Phaeodactylum tricornutum* and found that the polysaccharide chitosan (with pH adjustment) was more effective than either polyaluminium chloride (PAC) or aluminium sulphate (AS). It was effective even at concentrations as low as $20 mgL^{-1}$ provided the pH was adjusted to 9.9. The same diatom species was found by Spilling et al. (2010) to flocculate most effectively at a threshold pH of 10.5, achieved by turning off the CO_2 supply.

Leckey & Hinders, (2011) devised a method of acoustic manipulation for sorting and/or dewatering microalgal cells using standing waves. The method relies on the fact that particles in a standing wave field are pushed towards nodes or anti-nodes depending on their acoustic properties. The principle is illustrated in Figure 7, which shows that when a mixture of particles is passed through a channel on either side of which a transducer and reflector are precisely aligned, they come to occupy positions, either a pressure node or a pressure anti-node, as determined by their acoustic properties. The procedure can be applied to sorting desired cells (e.g. high lipid cells for biodiesel production or high starch content for ethanol production) from mixed cultures or from natural blooms. There is evidence that dewatering, a major cost factor in preparing microalgal biomass for further treatment, can be accomplished more efficiently and more economically by acoustic separation than by traditional drying methods (e.g. Araz et al. 2003). Other methods that have been used include ultrasound-assisted sedimentation, enhanced biosensor sensitivity and particle sorting (Bosma et al., 2003, Nilson et al. 2004, Kuznetsova & Coakley 2007).

Centrifugation requires high initial and continuing maintenance costs of equipment as well as high energy inputs. Filtration is also costly and best suited for larger species. Flocculation using metal salts or cationic polymers inevitably introduces contaminants and may affect the quality of the final product.

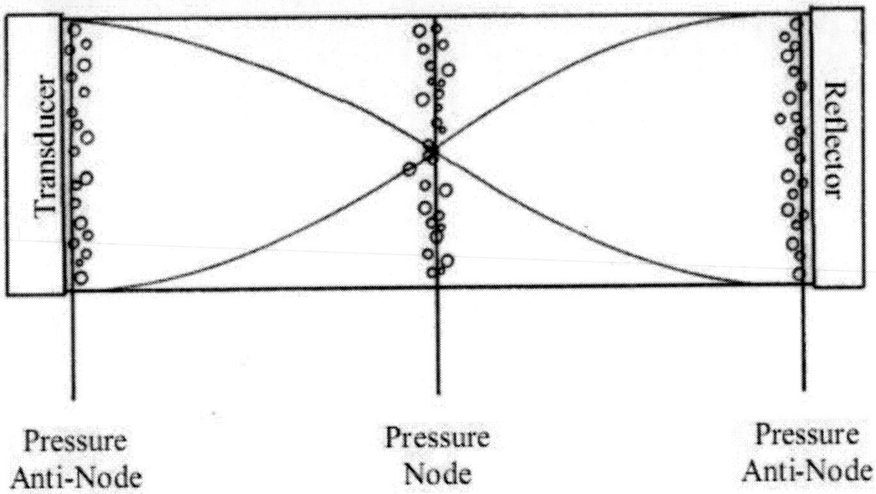

Figure 7. An acoustic particle separator adapted by Leckey & Hinders (2011) from Petersson et al., (2005) for harvesting and dewatering mass-cultured microalgae. Particles move to pressure nodes or anti-nodes in a standing wave field depending on their acoustic properties. (Redrawn, with permission, from Leckey & Hinders 2011).

Bioflocculation techniques, well known to be effective in a range of industrial processes including industrial fermentation and waste water treatment, are therefore increasingly being applied to the harvesting of microalgal biomass. The technique is based on the use of extracellular bacterial products (exopolysaccharides) to increase adhesion between microalgal cells and so cause aggregation. Lee et al. (2009) explored the possibility of using bacterial cultures to induce *in situ* flocculation of microalgae prior to harvesting. In their study using the marine microalga *Pleurochrysis carterae*, they stimulated the growth of the flocculating bacteria through the addition of organic substrates and then imposed nutrient depletion to induce excretion of polymeric substances. Using this procedure, microalgal recovery efficiencies of over 90% were achieved with a concentration factor of over 200. Vandamme et al. (2010) evaluated the potential of cationic starch for harvesting mass cultured *Chlorella* cells, and found that, under optimum conditions, 95% flocculation could be achieved without any observable toxic effects on the microalga. Moreover, unlike some other flocculants such as chitosan, cationic starch was effective at all pHs from 4 to 10.

The choice of harvesting procedure will depend on the microalgal species (or mixed population) and on the use to which the biomass is to be put. For human consumption, for example, the product is usually required in powder form which may require cell disruption prior to spray drying (Mohn, 1978). Microalgal populations used in waste-water treatment plants (see Chapter X) can be harvested by flocculation, followed by sedimentation and decanting for subsequent use or disposal.

Extraction of high value compounds from microalgal biomass has traditionally involved a range of cell disruption techniques and extraction with a range of organic solvents. Increasing use of the products in the nutraceutical industry has demanded more efficient and more environmentally acceptable procedures such as pressurized liquid extraction (PLE), which uses less solvent, is quicker and is more easily automated. Such PLE techniques have been used to extract the valuable carotenoid zeaxanthin from *Chlorella* (Koo et al. 2012) and for obtaining extracts with antiviral properties, as tested against Herpes simplex virus type 1 (HSV-1) from *Dunaliella salina* and *Haematococcus pluvialis* (Santoyo et al. 2012).

Chapter V

Microalgae as a Food Source

The sustainability of marine and freshwater fisheries which provide an important source of protein for human consumption world-wide (Garcia & Rosenberg 2010), is intimately linked with the functioning of the ecosystems that support them. Such ecosystem processes have been described as being under the influence of a triad of interlinked drivers: (1) exploitative, (2) biophysical and (3) trophodynamic (see Figure 8), in all of which the microalgal component plays a crucial role. Evidence for the strong link between fisheries stocks and the availability of microalgal primary producers has long been well known (see Smayda, 1991). Conditions that allow good growth of phytoplankton (e.g. nutrient input from upwelling or coastal run off) almost always lead to increased fish stocks. Dinoflagellates, a major component of many such microalgal "blooms" constitute an important food source for a number of fish species in natural waters, especially during the larval stage.

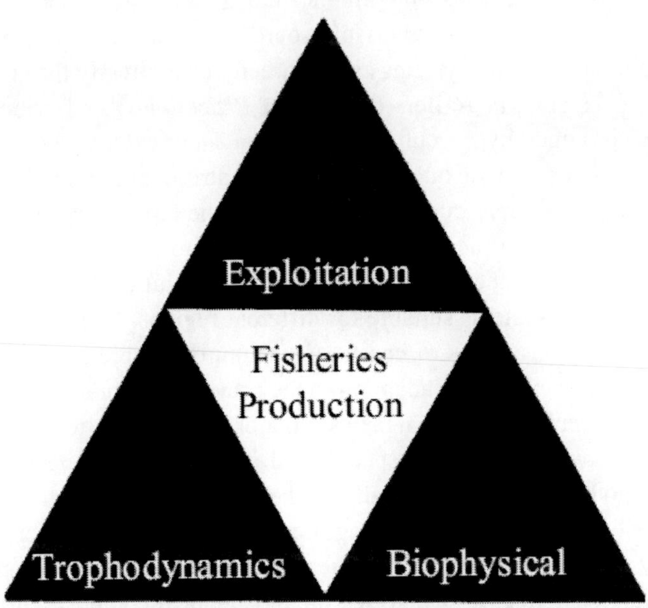

Figure 8. The triad of drivers affecting fisheries production. (Redrawn, with permission, from Link et al. 2012).

The stomach contents of the larval-stage of 20 North Sea fish species examined by Last (1980) were shown to contain significant amounts (up to 87%) of protozoan ciliates belonging to the Tintinnida (tintinnids), which are known to be major consumers of dinoflagellates, even while still in the yolk sac stage. Herring, sprat and sand lance larvae continue to ingest dinoflagellates until they are 8mm in length. The dinoflagellates, it was noted, were ingested year-round while diatoms were ingested primarily during the spring bloom.

The feeding of natural populations of fish larvae is critically affected by the availability of suitable concentrations (patches) of appropriate microalgal species. Lasker & Zweifel (1978) showed that the northern anchovy (*Engraulis mordax*) is attracted to dinoflagellate patches and stimulated to feed on the larger individuals if they occur in densities above a threshold level. Similarly, cod larvae (<6mm) primarily ingest tintinnids while mackerel larvae feed primarily on cladocerans (also major consumers of dinoflagellates), confirming the trophic linkages between fish stocks and microalgae via the microbial food web and zooplankton herbivory.

As a response to the well documented reports of dwindling stocks of many natural fish populations and of reduced catches due to overfishing, there has been an increased demand for alternative source of fish protein, a demand that can only be met by aquaculture.

Microalgae in Aquaculture

Microalgae are used as a food source in a range of aquaculture projects. They are consumed by bivalve molluscs (clams, mussels, oysters and scallops) throughout all their growth stages and by crustaceans and fin fish mostly during the early growth and larval stages, when they constitute an indispensable source of food. They may also contribute in a wide range of other aquaculture systems either directly or indirectly through being consumed by small invertebrates such as rotifers (species of *Brachionus*) and brine shrimp (*Artemia*) which in turn are consumed by the cultured organism. Their extensive use as a food source in a range of aquaculture projects is due partly to their nutritional composition and partly to the ease by which they can be captured and assimilated by the farmed organism. The likely future expansion of various forms of aquaculture will place increasing demands on the availability of suitable brood stock of the various species to be cultured, requiring the adoption of microalgal-based culture systems suitable for different regions. Particularly important will be the development of various forms of coastal aquaculture suitable for brackish environments using microalgal species capable of adapting to the prevailing conditions.

The suitability of cultured microalgae as a major food source for a range of aquaculture projects will vary between microalgal species and will also vary depending on the organism being cultured. Another variable, especially in the case of batch-cultured microalgae will be the state of the cells at the time of harvesting, that is, whether they are actively dividing (as in an exponentially growing culture) or whether they are in the post-exponential (stationary) phase of the culture. Whyte (1987) analysed the biochemical composition and energy content of batch cultures of two diatoms (*Thalassiosira pseudonana* and *Chaetoceros calcitrans*) and three phytoflagellates (two strains of *Isochrysis galbana* and *Tetraselmis suecica*) harvested at the exponential and stationary phase of growth. Based on a calorific evaluation of their lipid, carbohydrate and protein content, the microalgae were ranked in this study in the following decreasing order: *Isochrysis galbana* (both strains), *Chaetoceros calcitrans, Tetraselmis sueicica, Thalassiosira pseudonana* and generally, with the exception of the last named, the higher energy levels were recorded in stationary phase cells.

The highest concentrations of lipid were recorded in the *Isochrysis* strains while carbohydrates were higher in the diatoms. Total nitrogen values were higher in the phytoflagellates where protein formed a greater proportion of the organic matter, especially during the stationary phase.

The nutritional value of a number of microalgal species commonly used in aquaculture/mariculture, ranging in size from 6 to 20μm (and including some much larger diatom chains) was reported by Brown et al. (1997). Cultures were grown under controlled laboratory conditions and harvested towards the end of the period of exponential growth for analysis of their biochemical composition. The 40 microalgal species analysed differed markedly in their gross composition with protein values ranging from 6-52% of total dry weight, carbohydrate from 5-23% and lipid from 7-23%. Of particular relevance was the finding that diatoms, prymnesiophytes, cryptomonads and eustigmatophytes were rich in one or both of the key nutrients essential in animal nutrition, the polyunsaturated fatty acids 20:5 (n-3) and 22:6 (n-3) (5-35% of total fatty acids). Prasinophytes had low to moderate levels of one or other of these fatty acids (4-10%) whereas the chlorophytes were deficient in these essential components. It was concluded that since certain species were shown to be deficient in certain key nutrients (e.g. some specific vitamins), mixed diets are clearly preferable for most mariculture projects.

Microalgal species having a high content of polyunsaturated fatty acids (PUFAs) are particularly important for the culture of bivalves because most bivalves are unable to synthesise significant amounts of long-chain PUFAs (Delaunay et al .1993). The most common method for raising microalgal feed material for use in commercial bivalve culture is to grow the microalgae in plastic bags (500L) with 10% of the bag volume continuously harvested every day at late logarithmic phase. To confirm that this procedure produced feed with the required fatty acid content, Jacobsen et al. (2010) analysed harvested material from replicate bags containing cultures of *Isochrysis* sp.,*Pavlova lutheri* and *Chaetoceros muelleri.* They looked specifically at the essential fatty acids arachidonic acid, eicosapentaenoic acid, *n*-6 docosapentaenoic acid and docosahexaenoic acid and found that, despite considerable variability between bags, and some variation between species, the procedure yielded a stable supply of all the essential PUFAs to scallop larvae and spat from the three microalgal species when supplied in the ratio 1:1:1.

The long-chain poly-unsaturated fatty acids (PUFAs) mentioned above are common constituents of fish and fish oils and are well-known to be important in human metabolism. They lower the level of triglycerides and very low density lipoprotein in serum. They also have anti-inflammatory effects and are beneficial to the immune system (Leaf & Weber 1988; Drevon 1990). The low incidence of coronary heart disease in Greenland Eskimos has been correlated with the high concentrations of these PUFAs in their largely fish diet (Bang et al. 1980; Connor & Connor 1986).

As a contribution to exploring the link between the beneficial effects of fish PUFAs and the microalgae at the base of the food chain, Ahlgren et al. (1992) analysed the fatty acids of selected freshwater phytoplankton.

They reported that the flagellates *Cryptomonas, Rhodomonas*, and *Peridinium* had a relatively high fat content and a high proportion of PUFAs with a high ratio of n-3/n-6 forms. Two of the cyanobacterial species examined, *Oscillatoria* and *Microcystis* also had relatively high n-3/n-6 ratios, whereas *Anabaena, Spirulina* and *Nostoc* had only traces of n-3 acids or none at all. The chlorophytes generally had a higher content of PUFAs than did the cyanobacteria but with a similar n-3/n-6 ratio. Interestingly, the fatty acid content was strongly affected by the conditions of growth. As a general rule, the PUFA concentration and the n-3/n-6 ration was highest in cultures growing exponentially i.e. under non-limiting conditions. Under stressed (light- or nutrient-limited) conditions there was an increased lipid content due mostly to an increase in saturated fatty acids but little change, or even a decrease, in the more desirable n-3 PUFAs. Microalgal species containing PUFAs of the n-3 type would therefore be a very important component of the diet of cultured fish since they represent what would be the ultimate source of the fatty acids essential for biomembrane structure and function in humans.

A number of cultured diatoms have been successfully used in aquaculture. Two genera, *Skeletonema* and *Chaetoceros* have been extensively used in the culture of the larvae of brackish water shrimp (Simon, 1978; Yousoff et al. 2001). The diatom *Navicula,* was reported by Al-Hasam et al. (1990) to be particularly suitable as a food source in the mariculture industry because of its small size, its fatty acid composition (dominated by palmitic [16:0], palmitoleic [16:1] and eicosapentaenoic [20:5] acids) and its tolerance of high salinities (up to 1.7 *M* Na Cl).

In addition to their principal role as an indispesible food source during certain critical stages of the growth of a range of cultured species, microalgae are also widely used to improve the appearance, and hence the marketability, of the final product. The flesh of a number of wild salmonids, for example, is coloured by the presence of carotenoids which are not synthesised by the fish themselves but comes from dietary inputs. The fact that many microalgal species are known to be particularly good sources of a range of carotenoids (see Chapter VI), has led to their being extensively used in aquaculture, replacing carotenoids from other sources, both natural and synthetic, formerly used.

The microalga *Chlorella vulgaris,* whose cells contain significant amounts of the carotenoids canthaxanthin and astaxanthin, has been successfully used as an ingredient in the diet of cultured rainbow trout (*Oncorhynchus mykiss*) (Gouveia et al. 1997). Compared with diets supplemented by a synthetic mix of these two carotenoids, diets supplemented by *C. vulgaris* biomass (to give an equivalent supply of carotenoids) were shown to provide muscle colouration adequate for commercial purposes and only slightly inferior to that provided by the synthetic carotenoid supplements. Moreover, there was some indication that carotenoid retention in the *Chlorella*-fed salmonid flesh was well within the range reported for other carotenoid sources.

Despite efforts over several decades to develop cost-effective artificial diets to replace microalgae as hatchery feeds, on-site microalgal production remains a critical element for most marine hatcheries. The search continues for microalgal species with improved nutritional suitability for specific purposes such as the mass rearing of copepods for certain fish larvae. It is likely that there will in future be greater centralisation of microalgal production at specialised mass-culture facilities with improved harvesting, treatment and storage of the final product for distribution to hatcheries (Brown, 2002).

Microalgal Food Sources for Different Cultured Species

For many cultured fish species the connection between productivity and the availability of a suitable microalgal food source is well established, with the larval stage generally regarded as the most critical phase. Microalgal food additions in the form of cultured cells of *Chlorella* have been successfully used to enhance the growth of cultured nibbler fish (*Girella punctata*) (Nakazoe et al. 1986) and juvenile olive flounder (*Paralichthys olivaceus*) (Koo et al. 2001). In the latter study, dietary supplementation with 2% of *Chlorella* powder (prepared by pulverising freeze-dried cultured biomass of *C. ellipsoidea*) improved fish growth after 9 weeks, when expressed as weight gain and as feeding efficiency. It was however noted that dietary supplementation with both 2% and 4% *Chlorella* powder decreased whole-body fat levels, an observation explained as possible hormonal regulation of lipid metabolism resulting from the microalgal supplementation.

The beneficial effects of microalgae in fish culture are best observed when they are used to support growth of zooplankton grazers such as rotifers or *Artemia* which, in turn, are consumed by fish larvae. The larval stages of Japanese flounder (*Paralichthus olivaceus*), for example, are usually fed on rotifers for the first two weeks and later on *Artemia* nauplii (Takahashi 1985) both of which feed on microalgae such as *Nannochloris oculata* (Hur et al. 1989) or *Chlorella* spp. (Seikai et al. 1986). The microalgal food source is deemed to be superior to alternative sources such as baker's yeast which lacks fatty acids of the HUFA n-3 series (Lubzens et al. 1985) essential for the larvae of many marine fish (Watanabe et al. 1983). In a comparison of the suitability of different growth media used for the culture of *Paralichthus olivaceus* larvae, Cabrera & Hur (2001) reported that of the diets tested, growth and survival of the larvae over a 14 day period after hatching was best in those treatments where the rotifer food source had been fed *Chlorella ellipsoidea.* Cultures containing rotifers fed on *Nannochloris oculata,* or on yeasts, on the other hand, did not support good larval growth and resulted in higher mortalities.

As an alternative to supplying live microalgae as a food source to support the early growth stages of a range of cultured organisms, many have used dried microalgal suspensions as a major component of the diet. In their study of the feeding and physiological energetics of Sydney rock oysters, Kesarcodi-Watson et al. (2001) used varying concentrations of spray-dried cells of the microalga *Spongiococcum excentricum* (marketed commercially) to provide, upon rehydration, between *ca.* 27,000 and 180,000 cells ml^{-1}. They found that dried and rehydrated microalgae were readily ingested and effectively digested by both juvenile and adult rock oysters. Significantly, although a number of specially prepared diets such as yeast or yeast-based products have been tested (e.g. Nell et al. 1996), they have generally been shown to have a lower nutritional value than microalgal mixtures. Refrigerated off-the-shelf preparations of microalgal concentrates have been shown to be far superior to such products (e.g. Nell & O'Connor, 1991) as also have microalgal preparations based on chemical flocculation at high pH followed by treatment with a polyelectrolyte solution to produce sedimented flocculates for storage as concentrates for up to 30 days at 4° C (Brown & Robert 2002). Concentrates prepared following the microalgal flocculation method devised by Brown & Robert (2002) were successfully used as a major dietary component (80%) for the culture of Pacific oysters (*Crassostrea gigas*). For larvae of initial size 74-82 μm, best growth was obtained with diets containing *Chaetoceros calcitrans* forma *pumilum* concentrate although the growth achieved was considerably less than that achieved with the equivalent live preparation of this and other microalgae tested. Flocculation was therefore recommended by the authors as a cost-effective alternative to centrifugation for on-site preparation of microalgal concentrates of high nutritional value.

Typically, most marine finfish larvae are initially fed on rotifers (*Brachionus* sp), then on *Artemia* before finally being weaned on artificial diets. The inclusion of live microalgae in the diet has, however, been widely reported as being of particular benefit in the larviculture stage of various species (Reitan et al. 1993, Kennedy et al. 1998). It has been suggested, for example, that ingestion of live microalgae may help in the development of an embryonic gut flora that is protected against pathogenic bacteria (Papandroulakis et al. 2002) or in various other aspects of larval feeding or digestion (Oie et al. 1997). Intensive larviculture of sea bream (*Sparus aurata*) was shown by Papandroulakis et al. (2002) to be greatly enhanced in media to which exponentially-growing cultures of *Chlorella minutissima* were added twice daily to maintain maximum microalgal concentrations of $752 \pm 200 \times 10^3$ cells.ml^{-1}. Phytoplankton addition resulted in 44±17% survival of the bream larvae after 20 days with individuals of 2.0±0.2 mg wet weight. In the absence of phytoplankton the corresponding values were 16±6% for survival and 1.1±0.2 mg for wet weight. The daily consumption rate (as mg. carbon consumed per mg.body carbon) was 0.7 day^{-1} for larvae supplied with phytoplankton and 1.0 – 1.9 day $^{-1}$ for those reared in clear water. Among the observed beneficial effects resulting from the presence of phytoplankton in the rearing medium was better preservation of the endogenous oil droplet reserve of the larvae, generally indicative of superior quality of the exosupplied food. Growth performance of the larvae was better in the presence of phytoplankton even after the endogenous oil droplet had been totally absorbed. Another beneficial effect attributed to the inclusion of phytoplankton in the medium was an effect on the efficiency of larval feeding whether the main food source was rotifers (as in early exogenous feeding) or *Artemia* nauplii later.

Many marine organisms such as crustaceans, shellfish and salmonids contain the ketocarotenoid astaxanthin although they appear to lack the ability to synthesise this pigment *de novo*. Instead, they acquire it from their diet, in particular from the phytoplankton component, a natural source of astaxanthin and of a number of other carotenoids like ß-carotene, a precursor of astaxanthin (see Chapter VI). For many species, the beneficial effects of live microalgal additions can also be achieved by the use of equivalent amounts of microalgal paste. Thus, growth and survival of the larvae of the tropical/subtropical fish species cobia (*Rachycentron canadum*) fed on commercially available concentrated microlagal paste was as good as that achieved with supplied live cells of *Nannochloropsis* sp. (Schwartz et al. 2008). Similarly, microalgal pastes (prepared from concentrated suspensions of *Nannochloropsis*) are widely used in the mass production of the rotifer *Brachionus plictalis* (Pfeiffer & Ludwig, 2007).

Green Water Technologies

Several of the most common freshwater fish produced in world-wide aquaculture are obligate planktivores and consume little if any prepared feed (such as aquafeed) (Xie, 1999).

Their main feed is so-called "natural feed" which is composed mostly of microalgae (phytoplankton) together with some bacteria, and other micro-organisms, hence the term "green water". The green water component of different aquaculture systems will vary in composition and in its mode of production. It is usually produced in impoundments fertilized in various ways. Green water is a low cost food source which, under optimum conditions, is biologically "managed" by the different feeding habits of the fish being cultured, so controlling the composition of the plankton and the overall water quality of the system (Neori, 2011).

The food conversion ratio (FCR) (i.e. the weight of live food consumed per weight of fish or shrimp produced) has been estimated for various green water systems. For silver carp a conservative value of 8 has been reported (Berday et al. 2005). Systems based on zooplanktivory, by comparison, require much higher consumption of primary production for a given fish production because of the loss of energy and organic matter by the additional trophic level (Neori 2011). Tilapia, another of the world's leading cultured fish species, mostly grown in commercial operations with heavy feeding, is naturally a microalgal feeder and it has been estimated that approximately 30% of global tilapia nutrition is attributable to green water "natural food". Similarly with a number of other cultured freshwater fish and shrimp species, even when heavily supplied with aquafeed, the contribution of green water microalgae is considerable and there is clearly potential for further green-water use and hence greater cost saving.

The substantial cost savings of green water-based systems, compared with systems based on the mass culture of microalgal monocultures, come from the relative simplicity of the green water impoundments and their reliance on the skill and experience of the aquaculturist to maintain operating efficiency. The maintenance of an empirically determined mixed culture of fish and shrimp species reduces production costs while at the same time providing biomitigation capacity through accompanying organic wastewater treatment capacity (see Chapter X).

Harmful Effects on Aquaculture of Nuisance Microalgal Blooms

The conditions under which most cultured fish and molluscs are reared, i.e. in caged enclosures or raft culture, are those which might be expected to favour the rapid growth of certain microalgal species to form "blooms". It is inevitable, therefore, that as coastal and other aquaculture projects have expanded, the incidence of microalgal blooms has increased, resulting in serious stock mortalities due either to anoxia or, when the microalgal species are themselves toxic, to direct fish kills. The impact of noxious phytoplankton blooms on the aquaculture industry over the period of its early global expansion (1970-1990) has been reviewed by Smayda (1991). Various mechanisms are described through which microalgal blooms can have significant negative impacts affecting fecundity, survival and recruitment of the cultured species as well as effects on mortality of first-feeding, juvenile and adult stages. The effects can be recognised as water column events, usually associated with microalgal toxin production or anoxic events resulting from excessive microalgal growth beyond that which is utilised by grazers. The microalgal species associated with such harmful and financially costly events in the North Sea and Scandinavian coastal waters include toxic species of various dinoflagellate genera such as *Pyrodinium, Gymnodinium, Gyrodinium, Emiliania, Prorocentrum* and *Alexandrium,* as well as a number of other genera such as *Prymnesium, Chrysochromulina, Aureococcus* and some cyanobacterial species such as *Nodularia spumigena.*

Many cultured fish larvae are particularly sensitive to toxic dinoflagellates compared, for example, with filter-feeding bivalves, whose tolerance of dinoflagellate toxin is many thousand times greater than that of fish and other vertebrates (White 1984). It is this capacity of post-larval shellfish to accumulate large quantities of toxin (paralytic shellfish poison, PSP) that makes them such a hazard for human health. Planktivorous fish exposed to PSP-producing dinoflagellates such as *Pyrodinium bahamense* var. *compressum* seem to accumulate very little of the toxin in their flesh. Instead, the dinoflagellate cells accumulate in the gut so that serious ciguatera poisoning will occur only when the fish are eaten whole.

Capelin (*Mallotus villolus*) suffered heavy mortality levels when exposed in culture to the toxin-producing dinoflagellates *Alexandrium tamarensis* at concentrations equivalent to those commonly met with in coastal blooms (Gosselin et al. 1989). Herring larvae, winter flounder, red sea bream and Japanese anchovy reacted similarly (reported by Smayda 1991). As the feeding behaviour of fish larvae changes to become less dependent on dinoflagellates but more on larger prey such as tintinnids, nauplii, copepods and cladocerans, the susceptibility to toxification often remains. This has been attributed to vectoring of the original dinoflagellate saxitoxin through the food chain (e.g. White 1977). Brevetoxin produced by blooms of *Gymnodinium breve* blooms has been reported to be responsible for massive fish kills in Florida (Roberts 1979) while toxins produced by a number of *Chattonella* species have been shown to have a devastating impact on fisheries stocks in the Seto Inland Sea, Japan (Okaishi 1989). Similarly serious impacts on cod, whiting, pollock and saithe stocks have been reported for coastal blooms of *Chrysochromulina polylepis*, *Prymnesium parvum* and a number of other toxic flagellates in Scandinavian waters (Dundas et al. 1989; Johnsen & Lien 1989).

There is evidence that containment of fish or shellfish in cage or raft enclosures greatly increases their susceptibility to the harmful effects of toxic microalgal species. It has been shown, for example, that under such conditions a remarkably small prey dose of the highly toxic dinoflagellate *Alexandrium tamarense* can be lethal to first-feeding sea bream larvae (White et al. 1989). The relative insensitivity of shellfish to microalgal toxins only serves to prolong the potential harmful effects within the food chain. In noting the significance of dinoflagellates in the observed increased frequency, intensity and duration of noxious microalgal blooms, Smayda (1991) has concluded that the trophic level components and their feeding strategies probably reflect conditions that existed prior to the evolutionary appearance of the diatoms, when flagellates were the major primary producers. The remarkable nutritional dependency of first-feeding larvae on dinoflagellates is, according to Smayda, a further reflection of this evolutionary background.

The haptophyte *Prymnesium parvum* is another microalga known to cause extensive mortalities of cultured fish in many parts of the world (Guo et al. 1996). Among the control strategies adopted in various hatcheries, with varying degrees of success, are chemical treatments with ammonium sulphate (Barkoh et al. 2003) or copper sulphate. Treatment with probiotic systems or with barley straw, known to be effective in controlling some cyanobacterial blooms, has met with scant success in countering the damaging effects of this haptophyte in fish production ponds (Barkoh et al. 2008). Some synthetic bacillamides (e.g. 5-iodo-Bacillamide) have been shown to be effective in inhibiting the growth of some nuisance microalgae. The findings of Churro et al. (2009) indicate that a number of toxic cyanobacterial species were particularly sensitive as also were the diatoms tested. Chlorophytes and some non-toxic cyanobacteria, on the other hand, were less sensitive.

Microalgae As a Food Source for Direct Human Consumption and As Stock Feed

The number of microalgal species and the amount of microalgal biomass used directly by humans as a food source is growing at an unprecedented rate. The nutritional value of microalgae (especially the cyanobacteria) and their use as dietary supplements has been widely reported (e.g. Li & Qi 1997; Abdulqader et al. 2000; Spolaore et al. 2006). *Spirulina* (*Arthrospira platensis*), which was first mass cultured for commercial production in the 1980s (Voushak & Richmond 1988), is now highly valued both as a protein supplement and as a provider of pharmaceutical products (Richmond, 1988). It is widely cultivated in many countries, with China and the USA as major producers. The summary prepared by Liang et al. (2004) outlines the tremendous increase over recent decades in the use of microalgae as a source of human food or food additives in China, reflecting parallel advances in microalgal biotechnology in other parts of the world. Among the common foods having a small but nutritionally significant microalgal component, they list microalgal noodles, where the flour base is enhanced by the addition of nutrient-rich *Chlorella* or *Spirulina* powder (up to 1%). Microalgal bread or buiscuits have been described as containing enhanced vitamins, microelements and bioactive compounds. Soft drinks containing microalgal extracts have been marketed and it is reported that the addition of microalgal powder to green tea enhances its nutritional value, improves its dispersion characteristics and reduces irritation caused by caffeine. Microalgal tablets, capsules and nutritional drinks made from *Spirulina* powder are prized for their high amino acid content and are known to be particularly rich in vitamin B_{12} (with vitamins B_1 and B_2), carotenoids and xanthophyll pigments (Richmond, 1988).

The major use of *Spirulina* by humans is in the form of various health foods, especially if there is an already well established and stable market. Beneficial effects include those on intestinal flora (Ciferri, 1983) and on serum cholesterol levels (Gonzalez de Rivera et al. 1993). Successful therapeutic applications of *Spirulina* products to test animals include their use in cancer treatments (Ayehunie et al. 1996), in a range of viral infections including HIV-1, herpes simplex and, in humans, in the treatment of radiation sickness (Ciferri 1983).

Large-scale industrial production of *Spirulina* has been primarily located in tropical and sub-tropical regions of the world where the prevailing temperatures and abundant sunlight allow rapid rates of growth in raceway ponds or other mass culture facilities. Richmond & Grobbelaar (1986) reported on some of the factors limiting productivity of outdoor mass culture of *Spirulina* highlighting the effect of the light regime as affected by the depth of the open raceways and by the cell density of the culture. Later, Richmond et al. (1990) were able to show that in their subtropical, Mediterranean location, air temperatures during daylight hours exerted a predominant effect on productivity and that, in summer, photoinhibition was a major limiting factor, with growth ceasing almost entirely after mid-day. They concluded that in their location, it should be possible to achieve an annual output of some 60 to 70 t dry matter ha^{-1} yr^{-1} provided an optimal temperature was maintained and photoinhibition could be curtailed or even eliminated. The optimum time of harvesting (especially in summer) was in the evening, before the inevitable night-time loss of dry matter. Sustained output rates of some 25-27 g DW m^{-2} d^{-1} were predicted for the summer months (May to October) and some 12 – 14 g m^{-2} d^{-1} for the winter months (November to April).

In recent years there has been a significant development of a *Spirulina* industry in the Inner Mongolian region of China which has a mid temperate climate with an annual average temperature of only 6.4° C. (Lu et al. 2011). Despite this low temperature, commercial production of *Spirulina* biomass and *Spirulina* products has been successfully established in the region. Among the factors contributing to this success is the ready availability of suitable water in the form of numerous alkaline lakes (carbonate and bicarbonate concentrations of 260 mM or over) and underground water, the low average annual precipitation restricted to a short rainy season, an average 3,000h sunshine per annum (total annual solar radiation of 5,730-5,930 MJ m^{-2}), and the availability of extensive areas of land otherwise unsuitable for traditional agriculture. Some of the disadvantages, such as the low temperature and the frequency of strong winds and sandstorms have been overcome by the construction of plastic covers over the raceway ponds thus significantly raising the maximum daily temperature within the culture and reducing the diurnal temperature variation. Periods of rain or of particularly low temperature require special protection of the ponds by the additional of a thermo-proof cotton layer. The use of paddle wheels to maintain flow within the raceways, careful monitoring of various parameters within the culture and the adoption of appropriate techniques for biomass harvesting and reuse of old medium have all contributed to keeping down the costs and maintaining the quality of the product for human consumption, for animal feed and as a source of high-value products. It was estimated by Yamaguchi (1997) that up to 30% of the world's harvested microalgal biomass was, at that time, used as fodder additives, with a large proportion of this contributed by *Spirulina*. In particular, addition of up to 10% of microalgal material to the diet can provide the necessary protein requirement for healthy growth of poultry. Similarly, inclusion of *Spirulina* (5-20%) in the diet can improve the appearance and marketability of the flesh of various fish species (Spolaore et al. 2006). Of the trace elements essential for healthy growth and development in humans and other animals, selenium (Se) is particularly important. A number of degenerative conditions in humans have been attributed to Se deficiency (Schrauzer 2000; Surai 2006) caused by the generally low concentration of this element in most soils and consequently in animal feed. Se becomes specifically bound in selenoproteins as selenocysteine and selenomethionine and in that form is available for animal use whereas inorganic forms such as selenite (SeO_3^{2-}) and selenate (SeO_4^{2-}) are not. Some microalgae, like some other microorganisms, are able to incorporate Se from selenite into their biomass to yield a product enriched in Se for use as an additive in the diet of farm animals (Doucha et al. 2009).

Quality Control

Any microalgal product that is to be used in bulk for human or animal consumption would have to meet strict health standards. Pure cultures of *Spirulina,* for example, or extracts from such cultures, have been subjected to numerous safety evaluations without revealing any significant toxic effects. Hutadilok-Towatana et al. (2010) have confirmed the absence of acute and subchronic toxicities in laboratory rats fed with methanol extracts of *Arthrospira platensis*. Daily oral administration of doses up to 24mg kg^{-1} body weight over a 12 week period produced no observable abnormalities of behaviour, food and water intake or health status. Postmortem examination of the test group showed no gross or histological abnormalities. Although there have been few, if any, direct human toxicological studies, *Spirulina* has already been granted GRAS (Generally Recognised As Safe) status in the USA (Gershwin & Belay 2008).

In the case of cyanobacterial products, there is also the need to ensure that the cultured material does not contain as contaminants any of the many known toxic strains. It is known that under certain environmental conditions, a number of cyanobacterial species may produce toxic products that can be hepatotoxic, neurotoxic, cytotoxic or allergic (Chorus & Bartram 1999). Some others may contain tumour promoters (Falkoner 2005). Such toxic strains are often very difficult to distinguish morphologically from non-toxic strains. Natural blooms of the cyanobacterium *Aphanizomenon flos-aquae* harvested from Klamath Lake (USA), for example, are routinely subjected to mouse bioassays, or appropriate biochemical toxicity tests, before processing and marketing for human consumption (Carmichael et al. 2000). The application of molecular techniques is another safeguard that would ensure that a given product actually contains the claimed cyanobacterial spiecies/strain (Saker et al. 2007).

In their examination of 39 commercially available dietary supplements of cyanobacterial source, Rellán et al. (2009) tested for the presence of the neurotoxin anatoxin-a in several species of cyanobacteria. Three of the product samples tested (two labelled "Spirulina" and one labelled "Fish food") were found to be positive for anatoxin-a (at concentrations ranging from 2.50-33$\mu g.g^{-1}$). All the other products (most of them labelled "Spirulina" and three labelled "Aphanizomenon flos-aquae") yielded negative results. These results confirm previous reports that *Spirulina* is not known to produce anatoxin a and suggest that the low positive (but clearly detectable) toxicity reported in the Rellán et al. study may be due to contamination by other cyanobacterial species. That such contamination does often occur is evident from the work of Bruno et al. (2006) and Saker et al. (2007), the latter study detecting *Microcystis* (and the toxin mirocystin) in a product labelled as containing only *Aphanizomenon*.

Microalgae in Traditional Agriculture

Cyanobacteria and Rice Crops

It has long been a common practice to use nitrogen fixing cyanobacteria to improve nitrogen supply, and hence the yield of grain, for rice crops. The usual method is to inoculate paddy soils with heterocystous cyanobacteria previously grown in open ponds, and it has been estimated that yields equivalent to those obtained from the use of industrially produced fertilizers can be obtained at only a fraction of the cost (Roger & Kulasooriya 1980). An alternative method is to use the water fern *Azolla,* which contains the nitrogen fixing *Anabaena azollae* within cavities in its leaves (Lumpkin & Plunckett, 1980). The conditions under which these microalgal additions are most effective, and the methods used in assessing the beneficial effects, have been reviewed by Metting (1988). The increased yields obtained from *Azolla* additions are reported to be partly due to the nitrogen fixing capacities of the cyanobacterial symbiont and partly to the role of the fern as a green manure or companion crop. One of the standard methods for adding the *Anabaena-Azolla* association to rice crops is to add a large inoculum with the irrigation water leaving the *Azolla* to grow for a few weeks, then partially draining the field and ploughing the *Azolla* into the soil before planting. In this way, most of the nitrogen fixed by the cyanobacterial symbiont eventually reaches the soil or the flood water and thus becomes available to the rice crop (Kannaiyan, 1993).

Beneficial effects of the addition of dried cyanobacteria to rice crops were first reported by Watanabe et al. (1951). They recorded a 25% increase in yield following inoculation of a poorly drained paddy with *Tolypothrix tenuis.* A number of similar reports from various parts of the world have been reviewed by Whitton & Potts (2000), most describing positive results but some described as "equivocal". Others have described the effects of cyanobacterial addition to rice crops as "erratic" and "limited" (Roger, 1991).

The most commonly used cyanobacterial inocula occur as pure or mixed cultures of *Anabaena, Nostoc* and a few other nitrogen-fixing species. Standard protocols have been devised for *Nostoc* inoculations to ensure maximum delivery to the rice field of heterocysts (the nitrogen-fixing component of the cyanobacterial filament) and akinetes (perennating bodies resistant to adverse conditions). Innok et al. (2009) cultured the *Nostoc* sp. VICCR1-1 strain under conditions that induced the formation of akinetes and heterocysts. Omitting phosphorus and iron from the culture medium increased the number of akinetes by up to 20%, while absence of $CaCl_2$ induced a 30-46% increase in heterocyst formation. It was found that using *Nostoc* akinetes (mixed with montmorillonite clay) as inoculum, produced grain yields comparable with those obtained with chemical fertilizer applications. Moreover, the akinetes survived in the soil after harvesting, resulting in enhanced grain yield for both first and second crops, confirming the findings of Mishra & Pabbi (2004) that cyanobacteria can become established permanently if inocululation is done consecutively for three to four cropping seasons. Another technique that has been tried to give the inoculated strains (with especially high rates of nitrogenase activity) a competitive advantage under flood conditions is to immobilize the microalga on polyurethane foam (Kannaiyan et al. 1997). A field experiment showed that foam-immobilized *Anabaena azollae* excreted ammonia into the flood-water resulting in increased yields of rice grain and straw. Studies with liquid and soil microcosms in which chlorophyll (as a growth index) and nitrogen-fixing potential were measured showed that inoculation with paddy rhizosphere cyanobacteria resulted in better establishment and hence greater potential for improved crop yield (Prasanna et al. 2011).

Another potential contribution of cyanobacteria to the nutrient status, and hence the yield of the rice crop, comes from their phosphatase activity - their ability to release phosphate from a variety of phosphorus compounds (Banerjee & John 2005). It has been estimated that more than half of the filterable total phosphate in waters of deep-water rice fields may be in the form of dissolved organic phosphate and hence unavailable to the crop. Under such conditions, cyanobacterial phosphatases may be critically important in providing a source of phosphate additional to the, often limiting, supply of inorganic P. Banerjee (2007), in a comparative study of phosphatase activity of four rice field cyanobacterial strains noted wide variation between strains in the induction and activity of the two phosphatase enzymes, phosphomonoesterase (PMEase) and phosphodiesterase (PDEase). One strain (182), in particular, showed significant phosphatase activity even when exposed to high inorganic P concentrations.

Other Soil Microalgae

Microalgal additions to agricultural crops may have beneficial effects other than those due to nitrogen-fixing cyanobacteria. The use of microalgae to improve soil structure has been well documented (e.g. Bailey et al. 1973). Of particular interest in this regard are the palmelloid green microalgae, particularly various species/strains of *Chlamydomonas*, which, during cell division, may pass through a palmelloid stage when the daughter cells remain together, encased in a massive extracellular sheath of mucilage. It has been estimated that in some species of *Chlamydomonas*, the polysaccharide component of the mucilaginous sheaths can represent as much as 75% of the microalgal dry weight (Barclay & Lewin 1986). The beneficial effects of the application of such microalgal matter is thought to be due to the presence of certain acidic residues which enhance soil aggregation by providing cation bridges between negatively charged mineral particles. Application of palmelloid *Chlamydomonas mexicana* to irrigated soils in eastern Washington, USA, has been shown to increase significantly both aggregation and water retention (Metting & Rayburn 1983), with the potential for improved crop yield, reduced erosion and greater water-use efficiency.

Improving Crop Photosyntheis

Most of the world's major agricultural crops (rice, wheat, barley, canola, soybean) as well as the world's tuber and vegetable crops, are C3 species. Under present-day atmospheric conditions of low CO_2 and high O_2, the photosynthetic CO_2-fixing enzyme (RUBISCO) of these plants functions under conditions very different from the high CO_2 and low O_2 that existed 3.5 billion years ago when the enzyme first evolved (Badger et al. 1998). To cope with these environmental changes, modern C3 plants have adopted certain strategies for maximising the rate of carboxylation at these ambient levels of limiting CO_2. These include an improvement of the kinetic properties of RUBISCO such that the Km(CO_2) has decreased and the ability to distinguish against O_2 has increased at the expense of catalytic rate (Badger et al. 1998). Additionally, many C3 plants have developed active CO_2 –concentrating mechanisms (CCM) to improve CO_2-supply to RUBISCO at minimal metabolic cost. Despite these strategies, however, typical C3 species invest up to 30% of their soluble protein as Rubisco protein.

A possible alternative strategy would be to take advantage of the extremely efficient CCM possessed by many microalgae. Some cyanobacteria, for example, have the ability to concentrate CO_2 up to a 1,000 fold around the Rubisco protein, leading Price et al. (2011) to suggest that it might be possible to establish some form of the cyanobacterial CCM within the C3 chloroplast. This might involve transfer of functionally active cyanobacterial HCO_3^- transporters to the inner envelope membrane and transfer of a CO_2-uptake system to the thylakoid membranes. It might also reqire the establishment of a RUBISCO microcompartment like the carboxysomes of cyanobacteria. There is already evidence (Long et al. 2010) that engineering the assembly of a carboxysome within the chloroplast may be approaching feasibility. Moreover, as Price et al. (2011) have suggested, introducing an active cyanobacterial CCM into the C3 chloroplast, might result in other photosynthetic enhancement beyond the mere addition of bicarbonate transporters.

Chapter VI

High Value Products and Food Supplements from Microalgae

There is a long and ever-gowing list of compounds of microalgal origin now recognised for their potential as contributors to human health and nutrition or in the development of a range of other products of commercial interest. The list includes bulk essential oils (supplementing more traditional plant sources such as soybean, palm, sunflower or rapeseed) as well as a number of compounds of more specific metabolic importance such as antioxidants, vitamins, antibiotics or toxins. The suitability of microalgae as a potential source of commercial oil products has been known since the early observations of Spoehr & Milner (1949) who noted that the microalga, *Chlorella* could, under appropriate conditions, accumulate oil up to or even in excess of 70% of biomass weight. The chemical profile of a number of microalgal species, with emphasis on their lipid and hydrocarbon content, has been prepared by Ben-Amotz et al. (1985). This survey highlighted the variability between species in the quantity and composition of the compounds produced. There was also considerable variation with the growth conditions, especially nitrogen supply. Total lipid content as a percentage of organic weight ranged from a high value of 55% in *Botryococcus* grown under nitrogen deficiency to average maximum values of 25% in all the other species tested. A comprehensive listing of the lipid content of a wide range of cultured microalgal species has been prepared more recently by Varfolomeev & Wasserman (2011). Some of the highest values (in excess of 70% of total dry matter) were obtained with two green microalgal species *Botryococcus* and *Dunaliella*, although a number of other species, including some strains of *Chlorella* and some diatoms had lipid values greater than 50% of dry matter. The green microalgae were also generally the more productive in terms of oil yield when grown in culture, with some values in excess of 100mg L^{-1} day^{-1}.

The commercial feasibility of using oils from microalgal sources as a replacement for palm oil in the manufacture of soaps and lotions is now under serious consideration and algal oil has already been approved by the European Union as a food ingredient. Many other microalgal products, some already in production, have now been recognised and their nutritional and other benefits established in clinical trials. These include food additives of various kinds as well as compounds imparting flavours or fragrances. The green flagellate *Dunaliella,* for example, has been shown to be a good source of the triglyceride glycerol, much used in the cosmetic and pharmaceutical industries (Phadal & Sing 2003). *Dunaliella* is also one of the most prolific sources of β-carotene, a product which, like a number of carotenoids, is much used as a colouring agent in fish aquaculture and in the poultry industry. The β-carotene produced by *Dunaliella salina* has been described as being ten times more active than the synthetic pigment for a variety of nutritional purposes for both humans and animals.

There is increasing recognition that certain microalgae contain natural dyes that can be used to substitute for synthetic dyes used in the pharmaceutical, cosmetic, food and even the textile industry. Carotenoids have been widely used to colour various food products such as orange juice and phycocyanin is a common constituent of a range of creams and lotions. Polysaccharides from microalgal sources are commonly used to form gels and for thickening various products (Matsui et al. 2003). In addition to the major components of protein (50-65%), lipids (5-10%) and carbohydrates (10-20%), the biomass of *Chlorella pyrenoidosa* also contains vitamin C (200-500 mg.kg $^{-1)}$ and vitamin A (120-300 mg.kg^{-1}). Many cyanobacterial species produce extracellular polysaccharides (Volk et al. 2007) having properties useful in various industrial applications such as in water treatment plants for the removal of heavy metals. Many are effective in stabilising emulsions and some have been reported to have antiviral properties (e.g. de Phillipis et al. 2000, Santoyo et al. 2012).

Some microalgal polysaccharides have immunostimulatory properties, others are effective in scavaging free-radicals or in reducing blood lipid levels (Suárez et al. 2005, 2006). Other health related effects claimed for a range of microalgal products include lowering cholesterol levels (Khan et al. 2005), prevention of artherosclerosis (Cherng & Shih, 2005) and antitumor action (Noda et al. 1996). Weaning albino rats fed with dried preparations of *Dunaliella tertiolecta* as the sole protein source exhibited lower plasma cholesterol (and other lipidemic metabolic effects) than were found in control groups (Fabregas et al. 1988). Microalgal products shown to have potent immunostimulatory activity in human monocyte/macrophage bioassays include "Immulina" and "Immunon" from the cyanobacteria *Spirulina platensis* and *Aphanizomenon flos-aquae*, respectively and "Immurella" from *Chlorella pyrenoidosa* (Pugh et al. 2001). These microalgal polysaccharides were reported as being between one hundred and one thousand times more active for *in vitro* monocyte activation than other polysaccharide preparations used clinically for cancer immunotherapy. Phycocyanin, a biliprotein isolated from the cyanobacterium *Spirulina*, was shown by Romay et al. (2000) to exert substantial anti-inflammatory effects in animal tests measured as control of prostaglandin E_2 levels or as phosopholipase A_2 activity. In a search for treatments to manage such metabolic disorders as diabetes mellitus, Khursheed et al. (2012) screened a number of strains of *Spirulina* and reported evidence of insulin-like antigen effects which were confirmed by electrophoretic mobility tests.

Antibacterial effects attributed to certain microalgal products, first claimed for "chlorellin" produced by certain speciesof *Chlorella* (Pratt 1944) have since been confirmed for a number of other microalgal products and shown to be effective against *Staphylococcus aureus* and *Bacillus subtilis* but, interestingly, not against *Escherichia coli* (Debro & Ward 1979). A survey of a wide range of marine and freshwater microalgae showed that many exhibited strong protease inhibition, affecting specific enzymes but, in particular, the enzyme papain (Cannell et al. 1988). Other indications of potential clinical applications have come from experiments testing the effects of microalgal extracts on animal muscle contraction (Villar et al. 1994) and on the central nervous system (Laguna et al. 1993a and b; Villar et al. 1992a and b; Villar et al. 1994; Suárez et al. 2005, 2006).

A number of microalgal species are now recognised as prolific sources of protective substances able to counter the harmful effects arising from the action of free radicals on living systems. Many free radicals occur spontaneously but others arise from exogenous sources such as UV radiation, ozone, sulphur dioxide or a range of stress-inducing factors (e.g. exposure to high temperature, light or salinity, nutrient deprivation or anoxia). Oxidative stress occurs when the balance between prooxidant and antioxidant factors favours the former, resulting in structural molecular damage to key cellular components caused by free radicals (Kehrer & Smith 1994). Defence against the damaging effects of free radicals can occur either through repair mechanisms or through preventative mechanisms, the latter involving a range of antioxidant molecules such as ascorbic acid, reduced glutathione or tocols (tocopherols and tocotrienol - lipid-soluble molecules consisting of a hydrophilic chromanol ring and a hydrophobic phytol chain, and known to be responsible for vitamin E activity). The protective function of tocopherols in human health has been attributed to their antioxidant properties, in particular, to their capacity to prevent the oxidation of unsaturated fatty acids in cell membranes (Munné-Bosch & Alegre 2002; Sattler et al. 2004). Other products of known health-related properties are the glycerolipids that constitute a major component of the thylakoid membranes of microalgae, as they do of all photosynthetic organisms. Their biological activities include anti-inflammatory, antitumor-promoting, haemolytic and antviral properties (Reshef et al. 1997). The photosynthetic glycerolipids of three strains of the diatom *Skeletonema* have been shown to have significantly higher content of the two acidic components sulfoquinovosyldiaglycerol (SQDG) and phosphatidylglycerol (PG) than is the case with higher plants (Yan et al. 2011). In this respect, the diatom strains resemble the cyanobacteria (Wada & Murata, 1989).

Another group of compounds of proven value as food additives are the carotenoids for which microalgae are a particularly prolific source. Carotenes (otherwise described as provitamin A) have been shown to be effective in protecting against free radicals produced, for example, in human skin cells exposed to UV radiation. The mechanism involves protection of membrane phospholipids (Krinsky & Johnson 2005). Carotenoids such as astaxanthin (AXT) and lutein are very important as food dyes and as feed additives in aquaculture and in poultry farming (Lorenz & Cysewski 2000). Some carotenoids have shown considerable promise as effective agents in the prevention of a variety of human diseases, including cancer (Demmig-Adams & Adams 2002) and in delaying some forms of chronic disease. The development of cataracts and macular degeneration are both reported to respond to treatment with lutein (Olmedilla et al. 2003) as also is the progression of early atherosclerosis (Dwyer et al. 2001).

AXT, in addition to its effectiveness in lessening the damaging effects of ultraviolet radiation, is reported to stimulate production of high-density lipoproteins and to enhance the immune system (Guerin et al. 2003). It is recognised as having potential clinical applications as an antioxidant and is already much used commercially as a pigment source in fish aquaculture and in the poultry industry. Other beneficial effects claimed for AXT include protection against inflammation and against cancer and age-related diseases (Lorenz & Cysewski 2000; Guerin & Huntley 2003). Immature Wistar female rats, for example, were shown to receive significant protection from chemically induced mammary carcinogenesis when fed on a diet enriched with AXT (Florin et al. 2011).

Microalgal Carotenoids and Other Antioxidants

Carotenoid Biochemistry and Biosynthesis

Carotenoids are oxygen-containing, strongly unsaturated hydrocarbons with terpene-like character. They are closely related in chemical structure and physiochemical behaviour to two naturally occurring yellow pigments, lycopene and carotene (C_{40} H_{56}). The molecular structure of ß-carotene, one of three carotene isomers (differing in the position of the double bonds in the hydrocarbon chain) is shown in Figure 9a.

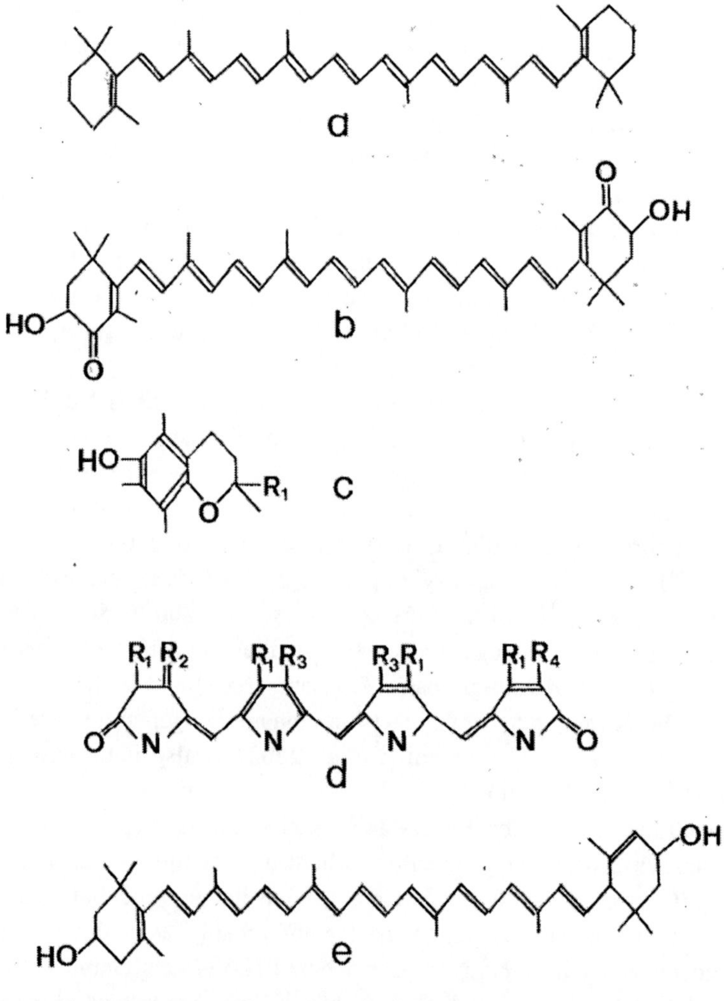

Figure 9. Summary molecular structures of some commercially important high-value compounds produced by microalgae. (a)-β-carotene, (b)-astaxanthin, (c)-α-tocopherol, $R_1 = C_{16}\,H_{33}$, (d)-phycocyanin, $R_1 = CH_3$, $R_2 = CH.CH_3$, $R_3 = C_2\,H_4.COOH$, $R_4 = C_2.H_5$, (e)-lutein. (Sources: Handbook of Chemistry & Physics, Weast, R. C. (Ed.) 59th Edition, 1978-79. CRC Press, Inc. Boca Raton, Florida, USA; Organic Chemistry, Karrer, P. 4th English Edition, 1950, Elsevier Publishing Co. Inc., Amsterdam; The Biology of Blue-Green Algae, Carr, N. G. & Whitton, B. A. (Eds.), 1973 Blackwell Scientific Publications, Oxford.).

It, like its isomer, lycopene, is readily oxidised by atmospheric oxygen, due to conjugation of the numerous carbon double bonds. Oxidation of the ß-carotene molecule produces a number of carotenoids in which the oxygen atoms occur largely as hydroxyl groups. One such carotenoid is astaxanthin (AXT) (see Figure 9b), which has four oxygen atoms, two associated with each of the terminal ß-ionone ring structures at either end of the hydrocarbon chain. AXT and lutein (Figure 9e) are synthesised from a common precursor (lycopene) through two divergent pathways with ß-carotene and α-carotene respectively as intermediates (Hirschberg 1999). Lycopene, the product of sequential denaturation of phytoene, is converted into ß-carotene by the action of lycopene ß-cyclase responsible for the cyclization of lycopene, an important branch point in the pathway of carotenoid synthesis (Sandmann 2001). Although carotenoids are produced in all photosynthetic organisms (plants, algae and cyanobacteria), and are also found in several non-photosynthetic bacteria and fungi, a range of microalgae have been shown to be particularly prolific sources of these isoprenoid pigments. They play an essential role in photosynthesis and nutrition as well as in protection of the organism against photo-oxidative damage.

The search for microalgal products having strong antioxidant properties has focussed particularly on vitamin E compounds. The enzymatic pathway of vitamin E synthesis in microalgae is now well known and the genes involved have been identified. Tocopherols, tocotrienols and plastoquinones, all compounds with vitamin E activity, are synthesised from homogentisic acid which, in turn, is synthesised from *p*-hydroxyphenyl pyruvate via the cytosolic *p*-hydroxyphenylpyruvate dioxygenase (Garcia-Plazaola & Becerril 1999). The other enzymes involved are located in the inner chloroplast membrane. One of them, γ-tocopherol methyltransferase (TMT), catalyzes methylation of δ- and γ-tocopherol to yield β- and α- tocopherol (Figure 9c). It has been functionally characterised in *Chlamydomonas reinhardtii* and its expression analysed (Gálves-Valdivieso et al. 2011). Expression of the gene coding for this enzyme (*tmt1*) is affected positively by high light and negatively by low temperature, confirming a probable role for γ- and α- tocophrol in adaptation of the microalga under low temperature and high-light stress conditions.

Preferred Microalgal Species/Strains for the Production of High-Value Compounds

Microalgal products marketed (as tablets, capsules or extracts) and advertised as being suitable for human consumption, have to be registered as having attained certain standards of purity and to have met the necessary food safety regulations. This will explain why a relatively small number of microalgal genera now dominate the market and why, all too often, the product standards may vary between different countries and between different suppliers. Attempts to guarantee the reproducibility of the health or other benefits claimed for different microalgal products are often hindered by a lack of precision on the taxonomy of the particular species or strain being used. This is well illustrated by the work of Görs et al. (2010) who investigated a range of commercial *Chlorella* products and their sources currently being marketed as dietary supplements for human consumption. Molecular analysis based on 18S rDNA indicated that most of the products contained more than one microalgal species. There was also evidence of bacterial or cyanobacterial contamination, raising concerns over the quality control standards applicable to some of the products being marketed. Such problems are often compounded by the fact that many manufacturers use their own production strains which have been found by trial and error to suit their particular cultivation techniques and downstream processing. Contamination by cyanobacteria is a particular concern because many of these organisms produce toxic compounds such as neurotoxins from species of *Anabaena* and *Aphanizomenon*, hepatotoxins from *Microcystis* and *Nodularia* species or dermatoxins from *Lyngbya* (see Chapter VII).

Despite the increasing recognition of microalgae as a viable alternative source of oils and of a range of other high-value compounds important in human nutrition, only a relatively small number of species have been fully investigated with respect to the yield and chemical characteristics of the desired product. Industrial production of polyunsaturated fatty acids (PUFAs), for example, is based largely on two genera, *Schizochytrium* and *Crypthecodinium* (Ward & Singh 2005) and scarcely more than a dozen or so species are in regular use for the production of carotenoids and other potentially valuable products. Selection of the most effective strains for commercial production involves careful isolation and screening, procedures that are both costly and time-consuming. As an aid in such selection programs, Guzmán et al. (2010) used the Nile Red staining method of Cooksey et al. (1987), in combination with flow cytometry (de la Jara et al. 2003), to estimate the unsaturation index and relative fatty acid content of cultures of four species of microalgae. The method is based on the observation that polar lipids, which are associated with membrane structures, generally have higher PUFA content than other lipids (Roessler 1990). Estimation of the ratio of polar to neutral lipids, as revealed by varying fluorescence with Red Nile (RN) stain, can therefore be used as an indirect indicator of relative PUFA content.

Haematococcus as a Source of Carotenoids

One microalgal species long known to be a good source of carotenoids is the chlorophyte *Haematococcus* (*Sphaerella*) (Volvocales, Family Sphaerellaceae) which occurs naturally in rock pools, bird baths etc and which, under conditions of high light exposure, accumulates enough pigment to colour the water blood-red or to settle as red granulations on surfaces (Proctor 1957). The potential of *Haematococcus* as a commercial source of the valuable carotenoid pigment astaxanthin (3,3'- dihydroxy-β,β-carotene-4,4'-dione) has been reviewed by Lorenz & Cysewski (2000). The application of special culture methods, they report, has allowed the mass production of *Haematococcus* containing 1.5-3.0% astaxanthin by dry weight. A one-step method for the continuous production of astaxanthin by cultured *Haematococcus pluvialis* has been reported by García-Malea et al. 2009. A photobioreactor operating continuously outdoors with nitrate input of 0.5 $mmol.g^{-1}d^{-1}$ yielded $0.7g.L^{-1}d^{-1}$ biomass and $8.0mg.L^{-1}d^{-1}$ astaxanthin.

The carotenoid is biosynthesized from acetyl–CoA and proceeds through the isoprenoid pathway via phytoene, lycopene, ß-carotene and canthaxanthin before the last oxidative step to astaxanthin. In green cells of *Haematococcus* the carotenoid fraction consists largely of lutein (75-80%) and ß-carotene (10-20%), whereas in the red cysts of the same microalga, astaxanthin can account for approx. 80% of the carotenoid fraction. Inclusion of *Haematococcus* in the diet has been reported to improve the colour of the shells of cultured shell fish and the tissues of cultured salmonids.

Accumulation of astaxanthin in *H. pluvialis* has been widely reported to be stimulated under various forms of environmental stress eg high light (Boussiba & Vonshak 1991), nutrient depletion (Borowitzka et al. 1991, Boussiba 2000) or high temperature (Tjahono et al. 1994). It has been suggested that astaxanthin, plays a major role in protecting the microalgal cell against photooxidative stress through shading of the photosynthetic machinery (Wang et al. 2003). A role for astaxanthin as an antioxidant against stress-induced excess reactive oxygen species (ROS) has been proposed (Kobayashi 2000). The relationship between irradiance levels, transcriptional expression of carotenoid genes and astaxanthin production has been studied by Li et al. (2010) using a wild-type and an astaxanthin-overproduction mutant of *H. pluvialis.* The results indicated that astaxanthin biosynthesis is mainly under transcriptional control of the gene encoding carotenoid hydroxylase and, to a lesser extent, the genes encoding isopentenyl isomerase and phytoene desaturase and, to the least extent, the genes encoding phytoene synthase and carotenoid oxygenase. Under elevated irradiance levels the expression of the plastid terminal oxidase was enhanced suggesting a role in controlling the formation of reactive oxygen species in the chloroplast and indicating more efficient photoprotective capability in the astaxanthin-overproduction mutant compared with the wild type.

Chlorella as a Source of Carotenoids

A number of different strains of *Chlorella* have been shown to produce carotenoids in potentially economically viable quantities. *C. zofingensis*, for example, when grown under high irradiance and nitrogen deprivation, accumulates high levels of astaxanthin and lutein (Rise et al. 1994; Bar et al. 1995). The ketocarotenoid astaxanthin can occur at concentrations up to 6.8 mg g^{-1} and in photoautotrophic batch culture at 7g dry weight l^{-1}(Del-Campo et al. 2004). Under heterotrophic conditions (in a glucose-supplemented medium), the yield of astaxanthin can reach values of 10.3 mg l^{-1} (Ip & Chen, 2005). Accumulation of total carotenoids in batch cultures of *C. zofingensis* was shown by Del-Campo et al. (2004) to reach a close-to maximum value of 40mgl^{-1} after 10 days, when the culture was well into its stationary phase. Accumulation of lutein, the dominant carotenoid, followed a trend analogous to that of the total carotenoids whilst astaxanthin, the second most dominant carotenoid, increased gradually over the entire culture period, reaching a maximal value at the end of the stationary phase (ca. 17 mg.l^{-1}, 1.7 mg.g^{-1} dry weight). Lutein was largely present in a free (unesterified) form whereas over 90% of the astaxanthin occurred as mono- or diesters. Within the range 20 – 28° C, greatest accumulation of both these carotenoids occurred at the higher temperatures. On a per volume basis, both lutein and astaxanthin values were higher at higher irradiance levels. Astaxanthin accumulation was higher under conditions of nitrogen deficiency which, however, supported reduced growth rates.

Huang et al. (2006) studied the biosynthesis of ketocarotenoids in *C. zofingensis* and identified and characterised a ß-carotene oxygenase (CRTO) gene directly involved in the process. Addition of glucose resulted in transient enhanced expression of the one *crtO gene* detected in the genome of *C. zofingensis.* A pathway for astaxanthin biosynthesis in *C. zofingensis* was proposed involving the participation of carotenoid ketolases which introduce a keto group to each of the two ß-ionone rings of ß-carotene and final conversion of the zeaxanthin formed to astaxanthin, catalysed by the ketolase 4,4'-ß-ionone ring oxygenase. It was concluded that biosynthesis of astaxanthin involved the initial formation of zeaxanthin *via* adonixanthin, which is then converted to astaxanthin by the *crtO* enzyme. There was also substantial production of canthaxanthin which accumulates because its further conversion to astaxanthin is limited by the conditions.

Dunaliella as a Source of Carotenoids

Dunaliella salina has been shown to accumulate ß-carotene (up to 10% of dry wt., or more) when maintained under growth limiting conditions such as high salinity, temperature and irradiance and/or limiting nutrients. Similar effects on intracellular accumulation of this compound have been observed in other *Dunaliella* strains (Vorst et al. 1994) suggesting a role for ß-carotene in light harvesting and photoprotection (Penuelas & Munne-Bosch 2005). Cultures of *Dunaliella salina,* exposed to salinities ranging from 0.6-4.5 M, showed best growth at NaCl concentrations within the range 1.5-3.0 M but were able to survive at salinities up to 4.5 M NaCl (Farhat et al. 2011). Carotenoid concentrations, however, increased with increasing salinity and, after the growth maximum had been reached, further salt addition (up to 4.5 M) led to enhanced carotenoid yield. Antioxidant activity (measured as gallic equivalents per unit dry weight), antiradical activity and iron-reducing power correlated well with carotenoid concentration. Borowitzka et al. (1990) studied the effect of sudden increases in salinity on growth and carotenogenesis in different isolates of *Dunaliella salina* and found that salinity up-shock (e.g. from 10% to 20% NaCl) resulted in an increase of total carotenoid (due mainly to β-carotene) and a decrease in lutein as a percentage of total carotenoids but an increase in zeaxanthin. The proportion of α-carotene remained unchanged.

Mass cultute of *Dunaliella* for commercial β-carotene production has traditionally been carried out in open ponds (Borowitzka 1992) but increasingly alternative methods are now being employed. Garcia-González et al. (2005) grew *D. salina* in a closed tubular photobioreactor maintained under batch and semi-continuous regime over a 40-day period. They found that under a batch regime the culture reached a maximum cell density of 6 x 10^9cells $L^{-1}$by the end of the first 9 days (see Figure 10a). At that point and at subsequent 4-day intervals the culture was diluted back to a cell density of *ca.* 2 x 10^9 cells L^1. This allowed maintenance of a specific growth rate of *ca.* 0.18 day^{-1} over the 40 days of operation. Maximum β-carotene content (up to 10% of dry biomass, $42mgL^{-1}$ culture) occurred at the end of the first 5-day period (see Figure 10b). Thereafter, during the semi-continuous regime, β-carotene production was maintained at between 20 and 30 mgL^{-1}confirming the potential of the closed tubular photobioreactor system as an alternative to the traditional open-pond system for the production of a high quality product for the health food market where the high operating costs might be more acceptable.

Regulation of the lycopene ß-cyclase (*LCY-ß)* gene of *D. salina* by abiotic stress has been reported by Ben-Amotz & Avron (1983). Gene expression and physiological analyses showed that steady state transcript of the gene and carotenoid levels were both enhanced in response to all stress conditions tested (salt, light and nutrient depletion). Nutrient availability, in particular, appears to be a key factor influencing carotenogenesis as well as carotenoid biosynthesis-related gene expression. High salinity and high light levels were shown to enhance the steady-state transcript levels of the two enzymes phytoene synthase and phytoene desaturase.

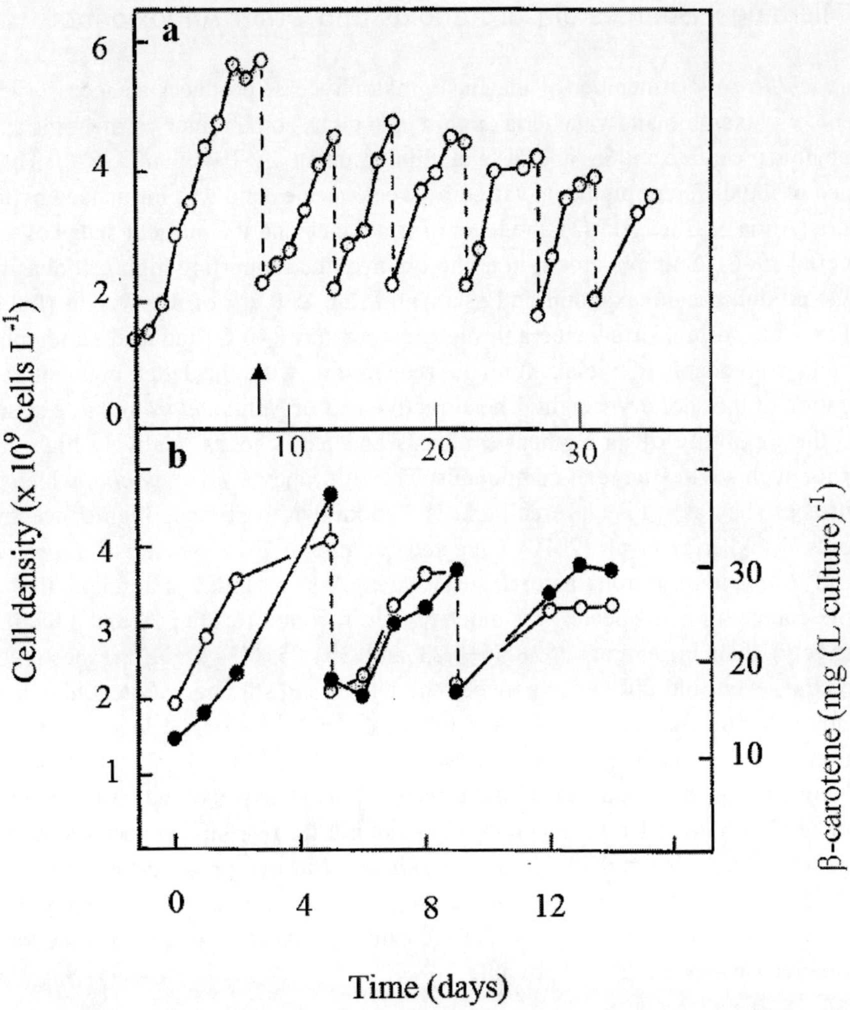

Figure 10. (a). Growth of *Dunaliella salina* in a closed tubular photobioreactor under batch (0 to *ca.*9days) and semi-continuous culture regime initiated by the first dilution (arrow) and maintained thereafter by periodic dilutions. (b). Biomass (open circles) and β-carotene (closed circles) accumulation in *D. salina* culture in a closed tubular photobioreactor under natural daylight conditions. Before the start of the semi-continuous culture shown, the culture had been maintained for six weeks under the culture conditions depicted in (a) (Redrawn, with permission, from García-González et al. 2005).

Ramos et al. (2008) isolated and charcterised the gene *Lcy-ß(DsLcy-ß)* which encodes for the enzyme responsible for the synthesis of β-carotene from the polypeptide acyclic lycopene (Lcy-β). Sequence similarity analysis revealed a close similarity with ß-cyclases from other plant and algal sources. The *Dunaliella salina* genome, they further showed, has only one copy of the *Lcy-ß* gene. They confirmed regulation of the lycopene ß-cyclase (*LCY-ß*) gene by abiotic stress. It was further concluded that the observed simultaneous regulation of genes encoding the first three enzymes of carotenoid biosynthesis is unlikely to be the only mechanism controlling this pathway and that post-transcriptional, translational and metabolic regulation may also be involved (Steinbrenner & Linden 2001).

Other Microalgal Sources of Carotenoids and Other Antioxidants

Nannochloropsis, a member of the Eustigmatophyceae, produces as accessory pigments ß-carotene, violaxanthin and vaucheriaxanthin as well as some minor carotenoids such as the ketocarotenoids canthaxanthin and astaxanthin (Lubián & Establier 1982). The relative abundance of the different pigments varied between species and was influenced by the age of the culture (Antia & Cheng 1982), the level of irradiance and the nutrient status of the culture (Lubián et al.1986). Cultures grown in a photobioreactor supporting high cell densities (~10^9 cells ml^{-1}) produced canthaxanthin and astaxanthin up to 0.7% of dry weight (Lubián et al. 2000). Exposure to high irradiance and high temperature (40°C) induced xanthophyll cycle activity and conversion of violaxanthin to zeaxanthin, with the latter pigment reaching a concentration of 0.6% of dry weight. The attractiveness of *Nannochloropsis* as a source is due partly to the flexibility of its pigment synthesis and partly to its relatively high production capacity for such a wide range of compounds. The chlorophyte *Trentepohlia,* which grows on moist surfaces such as walls and tree bark, is also known to produce high concentrations of carotenoids. Mukherjee et al. (2010) collected samples of two species of *Trentepohlia* (*T. aurea* and *T. cucullate*) from natural surfaces in Assam, India and found that the total carotenoid content of both species was much higher in winter (6.31±1.24 and 3.88±0.41 mg g^{-1} respectively) than in summer (838.56±0.23 and 341.05±0.34 µg g^{-1} respectively). They attributed the seven-fold difference between the winter and summer carotenoid content to the high summer temperatures (30-40°C) and overcast conditions of the locality and the lower temperatures (ca. 25°C) and cloudless clear skies of winter.

Chlamydomonas has been shown to produce two major types of antioxidants, carotenoids and tocopherols (vitamin E). In an attempt to examine for overlap between the two, Li et al. (2012) genetically engineered strains of *C. reinhardtii* to overproduce tocopherols and then tested whether such overproduction could compensate for carotenoid deficiency in a mutant lacking zeaxanthin and lutein. The results indicated that such overlapping functions against photooxidative stress were indeed possible.

Cyanobacteria as Sources of Valuable Bioactive Compounds

In their review of the potential of cyanobacteria as sources of bioactive compounds, Prasanna et al. (2010) reported that these prokaryotes represent a hitherto almost completely untapped bioresource for a diverse range of products. Among the cyanobacterial products already in use in bioindustry are the phycobiliproteins (comprising the pigments phycocyanin, phycoerythrin and allophycocyanin) which may account for approximately 20% of total cyanobacterial dry weight. It is predicted that the nutraceutical and pharmaceutical properties of these compounds may find widespread industrial applications in the future. Cyanobacteria are also rich sources of terpenoid pigments such as ß-carotene, zeaxanthin, ketocarotenoid, echinenone and myxoxanthophyll, all with interesting properties.

Phycocyanin (see Figure 9d) and other pigments from *Spirulina* and *Aphanizomenon* are already in use as colour additives in food and cosmetic production and some have exhibited pharmacological properties in the treatment diseases such as Alzheimer's and Parkinson's (see review by McCarty, 2007). The phycocyanins are assembled, along with their related phycobiliproteins, in supramoleular aggregates (phycobilisomes) which occur in association with the thylakoid membranes. Extraction of phycocyanins involves disruption of the cells (by sonication, freeze-thawing or treatment with enzymes) and purification by ammonium sulphate precipitation and follow-up treatments. Since the phycobiliproteins are water-soluble, extensive purification is essential. Gupta & Sainis (2010) devised a single-step procedure to obtain fairly pure phycocyanin from a species of *Synechococcus* sp. (which does not contain phycoerythrin) with a yield of 80-100 $mg.g^{-1}$dry weight of cells. The method, which uses lysozyme and charcoal-chitosan treatment is fast and, according to the authors, suitable for large-scale production of this much sought-after product.

Among the other products produced by cyanobacteria are the phytohormones (auxins, gibberellins, cytokinins and abscissic acid) compounds that act in the control of various aspects of the germination, growth, flowering and seed-production of higher plants and hence of potential application in agriculture or horticulture. Gibberellin-like substances have been reported to be produced by *Phormidium foveolarum* (Gupta & Gupta 1972) and by *Cylindrospermum* spp. (Mohan & Muherji 1978). Many species of cyanobacteria are known to produce large quantities of extracellular polymeric substances (exopolysaccharides) some finding commercial application in improving soil conditions (Caire et al. 1997). These easily recoverable polysaccharides often show greater potential for industrial or pharmaceutical application than is the case for polymers extracted from other sources. Cultures of *Nostoc flagelliforme* were shown by Yu (2012) to release extracellular polysaccharides (especially when supplied with various organic sources) with applications in food, textile, paint industries or pharmaceutical industries.

Some benthic cyanobacterial genera such as *Lyngbya* and *Symploca* have been shown to be particularly rich sources of compounds belonging to the mixed polyketide-polypeptide structural class (Tan 2007). Many of them (e.g. rifamycin, lovastatin and vancomycin) have been used for the treatment of human diseases such as inflammation, bacterial infections, and cancer (Simmons & Gerwick 2008). A high proportion of these bioactive compounds have been shown to target the eukaryotic cytoskeleton, such as tubulin and actin micro-filaments, indicating their potential as anticancer drugs (Jordan & Wilson, 1998). One group of such microtubule disruptors are the dolastatins, linear pentapeptides capable of binding to tubulin, suggesting that they or their structural analogues might be effective as antitumor agents (Watanabe et al. 2007). Other cyclic lipopeptides such as those isolated from the cyanobacterium *Hormothamnion enteromorphoides* have also proved to be effective in treating cancer cells, especially when used in combination with other related analogues (Bonnard et al. 2007). Continuing biosynthetic studies have confirmed the remarkable versatility of cyanobacterial enzymic systems in generating unusual functional groups with strong cancer chemotherapy potential.

Among the range of bioactive secondary metabolites produced by cyanobacteria are some that have been shown to exhibit strong allelopathy i.e. general or more specific antibiotic effects on their immediate environment (Schlegel et al. 1999). Such effects may either elicit positive or, more often, negative (deleterious) effects on other organisms indicating potential therapeutic and agrochemical applications. One such cyanobacterial allelochemical is cyanobacterin, a chlorinated γ-lactone produced by *Scytonema hofmannii* which inhibits the growth of other microorganisms (Mason 1982). Another is fischerellin A (produced by *Fischerella muscicula*) which is an inhibitor of photosystem II (Hagmann & Jüttner 1996). Exudates from the freshwater cyanobacterium *Oscillatoria* have been shown to inhibit the growth of *Chlorella vulgaris*, with maximal allelopathic effect during the early growth stages of the cyanobacterium (Leão et al. 2010). The active ingredients, which seemed to act synergistically, were two cyclic peptides containing several unusually modified amino acids. Production of toxins by cyanobacteria and its biological and environmental significance is described in Chapter VII.

Techniques have now been applied to some cyanobacterial species to maximise the yield of the desired product by supplying suitable precursor molecules to the culture. Yield of the important anti-inflammatory unsaturated fatty acid γ-linolenic acid (GLA, 18:3) in cultures of *Spirulina platensis* has been shown to be greatly enhanced following the addition of long- or short-chain carbon sources (Kim et al. 2012). Using the two-stage culture technique (see Chapter VIII) quite high yields of GLA (2% w/w and a productivity of 27.6±4.7 mg L^{-1}) were obtained by using 0.01 mM palmitic acid as a supplement. Such techniques as this suggest that cyanobacterial cultures may provide an alternative source of GLA which has traditionally been sourced from seed oils or from several species of fungi. The latter, in particular, have higher production costs as well as other problems relating to taste and odour.

The significant contribution of diatoms to annual marine biomass production (up to 40%, according to Falkowski et al. 1998) is related to their production of a range of unsaturated omega-3 fatty acids such as eicosapentaenoic acid, arachidonic acid and docosahexaenoic acid, all potentially important ingredients of zooplankton diets (see Chapter V). Such compounds as these have also been shown to have a range of other attributes of more direct importance to man. They can reduce cardiovascular disease (Lebeau & Robert 2003), they are precursors for certain tissue hormones and are present in grey matter of the brain (Kroth 2007). Eicosapentaenoic acid (EPA) can account for 3.9-5% of the dry weight of autotrophically grown *Phaeodactylum tricornutum* with an average contribution of 118 $mgL^{-1}d^{-1}$ to the total omega-3-fatty acid complement (530mg.$L^{-1}d^{-1}$) (Meiser et al. 2004). Compared with fish liver oil, an alternative natural source of EPA, diatoms offer a less expensive and in many ways a more acceptable source of this valuable product.

Schizochytrium, a heterotrophic microalga (a thraustochytrid, belonging to the Labyrinthulomycetes, formerly classified as a fungus but now placed in the Stramenopila – Porter, 1990) is now recognised as a good commercial source of the polyunsaturated fatty-acid docosahexaenoic acid (DHA). Originally isolated from leaf litter in mangrove regions of coastal Vietnam (Hong et al. 2007), *Schizochytrium* is used (in the form of dried powder) as an ingredient for poultry food and in aquacultural applications, as a dietary source for shrimp larvae and various finfish. Hong et al. (2011) investigated the conditions required to maximise biomass production and DHA content of cultures of the microalga grown in a 50% seawater medium containing organic supplements. They found that after 5 days culture at optimal conditions (28°C, pH 7.0, salinity 1.5%, 30gL^{-1}glucose, 1% yeast extract) their isolate (strain PQ6) achieved a total lipid content of up to 38.67% of dry weight and 1.57g DHA per L of culture. Scaling up to 5 and 10L fermenters yielded a minimum lipid content of approx. 38% of microalgal dry weight, with more than 50% of that as DHA. It was concluded that *Schizochytrium* was a promising source of DHA, a fatty acid well known for its possible medical applications in assisting visual and neurological development in infants and in the mental health of adults (Willatts & Forsyth, 2000).

Factors Affecting Production of High-Value Products by Microalgae

It has been noted that in some, but not all cases, cultivation of microalgae under non-optimal conditions (e.g. nutrient limitation or high irradiance levels) appears to stimulate production of certain commericially valuable compounds. Among the different forms of stress that have been shown to stimulate the production of various compounds of potential commercial interest is high levels if incident radiation, especially UV-B irradiation which often results in an increased synthesis of protective compounds such as carotenoids (Farhat et al. 2011; Wang et al. 2003). Other form of stress such as exposure to high salinity can induce the synthesis of compatible osmolytes (such as glycerol in *Dunaliella*) as well as other compounds such as β-carotene (also in *Dunaliella*) (Oren, 2005). In their study of fatty acid production by three microalgal species (*Isochrysis galbana, Tetraselmis suecica* and *Dunaliella salina*), Guzmán et al (2010) confirmed that nitrogen starvation resulted in a decrease in the unsaturation index of the fatty acid product. With the other species tested (*Chlorella pyrenoidosa*) there was no such response to nitrogen starvation. The plasticity in fatty acid composition was primarily due to variation in PUFA and MUFA (poly- and monounsaturated fatty acids) content, whereas SATs (saturated fatty acids) showed high stability.

High levels of oil accumulation are often associated with nutrient limitation (N for most species, Si for diatoms) which, in most cases, also resulted in reduced productivity. Many microalgal species, however, have the capacity to supplement their autotrophic (photosynthetic) nutrition with heterotrophic nutrition based on the utilisation of supplied organic substrates as a source of carbon and energy. Under such conditions, and some other strictly defined conditions, lipid biosynthesis can be "triggered" without necessarily reducing overall productivity. It is suggested that in microalgae that accumulate high levels of oil, the rate of growth is lower than the intrinsic rate of lipid biosynthesis and the cells assimilate carbon quicker than they can convert it into new growth. The excess carbon is therefore converted to lipid (Ratledge & Wynn 2002).

Growth of *Chlorella* in outdoor photobioreactors supplied with enhanced CO_2 concentrations (derived from combustion of organic waste or from fermentation processes) has been shown to yield high levels of starch biomass (Doucha et al. 2005; Douskova et al. 2009). Brányiková et al. (2011) were able to confirm this using the strain CCALA 924 from the Třeboň (Czech Republic) collection, chosen because of its high growth rate (μ_{max}= 0.20 h^{-1}), its tolerance of high temperatures (up to 40°C), its resistance of shear stress, its low adhesion to the surface of the bioreactor vessel and its low tendency to form aggregates (Doucha & Lívanský,2009). Its ability to grow under high concentrations of CO_2 (up to 13% v/v, Douskova et al. 2009) opens up the possibility of using flue gas sourced from waste incinerators, power stations, lime kilns and other industrial processes (see Chapter XI below).

Brányiková et al. (2011) used synchronised batch cultures to monitor the growth and starch content of *Chlorella* cells over two cell cycles. They noted that as biomass density increased (and light intensity within the culture declined) the relative starch content gradually increased to reach a maximum of 46% of DW just before the first cell division (after about 10h). During cell division, the relative starch content decreased (even in cultures maintained in light) to 13% of DW. If cell division occurred in the dark, the stored starch was almost completely exhausted (to 4% of DW). In the presence of cycloheximide (a specific inhibitor of cytoplasmic protein synthesis), the pre-cell-division starch content was increased to 60% of DW. A similar response was achieved with sulphur limitation. Under phosphorus or nitrogen limitation, the maximum starch concentrations were 55% and 38% of DW respectively. Significantly, sulphur-limited and cycloheximide treated cells maintained their high starch content for up to two days. In an outdoor pilot-scale trial, sulphur-limited cultures attained a starch content of 50% of DW after 120h compared with ca. 18% for cultures in a complete mineral medium. Such high concentrations of starch would be particularly suitable as source material for biosynthesis of triglycerides and subsequent biodiesel production. Exceptionally high concentrations of starch have also been noted to accumulate in the much enlarged and structurally modified chloroplast of the so-called "giant" cells of the Emerson strain of *Chlorella* produced when cultures of this strain are maintained under purely heterotrophic conditions in the dark (Griffiths & Griffiths, 1969). The starch content of the "giant" cells can be over forty times (on a per cell basis) that of autotrophically grown cells (Griffiths, 1965).

The ability of microalgae to respond to different cultural conditions by varying the rate and level of accumulation of different cellular products points to their potential as providers of raw materials for the synthesis of many commercially useful macromolecules. The cyanobacterium *Nostoc muscorum,* for example, can accumulate the macromolecule poly-ß-hydroxybutyrate (PHB).to 21.5 % of the dry weight in CO_2 aerated, phosphate starved cultures (Haase et al. 2012). Other cyanobcteria such as *Spirulina platensis* and *Synechocystis* sp have also been shown to accumulate PHB, especially under nutrient-limiting conditions (Panda et al. 2005; Wu et al. 2001). PHB has a number of attractive features including its thermoplastic, non-toxic and crystalline nature, all of which would make it a potential alternative source of plastics, replacing those from fossil material. It is not branched, has a melting point of ca. 179°C (Booma et al. 1994) and, with suitable treatment, can form products similar to polypropylene and polyethylene suitable for packing materials and a range of other products (Reddy et al. 2003).

The Physiological Role of Carotenoids in Microalgae

Different carotenoids are known to have different metabolic functions. Lutein, for example, is generally considered to be a primary carotenoid because of its essential role as a component of the photosystems. It acts as an accessory pigment of light harvesting complexes and also serves a protective role against damage of the photosynthetic apparatus by excess solar energy or other stresses. Astaxanthin, on the other hand, is a typical secondary carotenoid which accumulates in lipid bodies located outside the chloroplast (Bar et al. 1995; Margalith 1999; Grunewald et al. 2001). Secondary carotenoids provide protection against UV-light photooxidation (Del-Campo et al. 2004), acting as antioxidants, preventing accumulation of oxygen radicals. Nitrogen starvation and high light irradiance induces a drop in chlorophyll and primary carotenoids (ß-carotene, lutein) levels in *Chlorella zofingensis* but greater accumulation of secondary carotenoids (astaxanthin, canthaxanthin) (Rise 1994; Bar et al. 1995). The carotenoid pool in *C. zofingensis* was shown by Del-Campo et al. (2004) to reach $40 mg.L^{-1}$ during the late stages of the stationary phase of batch cultures, with lutein and astaxanthin as the major components. The different responses of carotenoids to different culture stresses opens up the possibility for control over the accumulation of the preferred carotenoid. The cellular concentration of astaxanthin, for example was more responsive to stress conditions compared with lutein. The aim in culture studies would be to balance the stress effect on production of the desired compound against effects on overall growth of the culture.

Photoprotection in microalgae generally (and in higher plants) has been reported to involve two main processes, namely the so-called xanthophyll cycle and the damage-repair cycle of PSII reaction centres (Qiu & Li 2006). The former consists of a pH-dependent conversion of violaxanthin, first to antheraxanthin and then to zeaxanthin. The latter involves a D1 protein repair cycle effective in protecting a key protein in the reaction centre of PSII. This it does by replacing the oxidatively damaged and defective protein by a new protein copy. Both mechanisms have been shown to be active in *Haematococcus pluvialis,* with the D1 protein repair cycle as the most important despite the fact that the carotenoid content of cells exposed to high light (250 μmol m^{-2} s^{-1}) (red cells) was 435% of that of cells grown at 50 μmol m^{-2} s^{-1}(green cells). The operation of the D1 protein repair cycle was reported by Qiu & Li (2006) to alleviate photoinhibition by 49% in green cells and 53-55% in red cells. The role of the xanthophyll cycle in microalgae is somewhat ambiguous because its contribution to energy dissipation can vary significantly between species (Masojidek et al. 2004). In *H. pluvialis*, according to Qiu & Li (2006) the xanthophyll cycle could contribute to the protection of green cells subjected to strong irradiance but its role in red cells was negligible.

Marxen et al. (2010) grew cultures of the cyanobacterium *Synechocystis* under two continuous culture control systems: turbidoststic control in which the culture was exposed to constant UV-B irradiance and physiostatic control in which UV-B irradiance was modulated (by means of an electronic control loop) in such a way that photosynthetic efficiency was kept constant at a fixed point.

The physiostat was devised as a culture method aimed at achieving a balance between attaining maximum stimulation of synthesis of the desired substance and the harmful effects on the organism of excessive exposure to UV-B irradiance. The physiostat seeks to keep a physiological parameter constant (e.g photosynthetic yield as measured by chlorophyll fluorescence).

An electronic feed-back loop compares the measured photosynthetic yield with a given set point and adjusts the UV-B irradiation in such a way that the difference between set point and actual yield is kept at zero. The external electronic feedback loop over-rides the action of the cellular feedback loop which normally adjusts protective pigments and antioxidants to the actual damage. Marxen et al. (2010) measured UV-B-induced stimulation of the synthesis of pigments, α-tocopherol and also the antioxidant potential of methanolic soluble components. They noted significant differences depending on the mode of irradiation, even though the overall doses were the same. For example, compared with the initial values, the concentrations of myxoxanthophyll and zeaxanthin increased by >200% and >400% respectively under constant UV-B irradiation in turbidostatic culture. In cultures under additional physiostatic control the corresponding increases were 600% and >700%. For α-tocopherol the increases were 150% under constant UV-B irradiances and nearly400% in cultures under physiological control.

The superiority of the physiological control regime for the production of these compounds was explained as interference by the physiological controls of the transducer chains that would otherwsie adjusts the production of protective pigments. The authors devised a complex scheme of interactions and feed-back loops as a putative mechanism to explain the findings they observed. The scheme incorporates input UV-B as it affects both the antennae pigments and the pool of UV-B-induced radicals as well as the generation of protective agents scavenging the radicals. The observed changes in the concentrations of myoxanthin and zeaxanthin and α-tocopherol following exposure to UV-B irradiation indicates a dominant role for these agents in the protection of *Synechocystis* (and probably other cyanobacteria as well). The scheme devised indicates some of the factors that have to be taken into consideration in any attempt to devise culture conditions that maximise the yield of these commercially valuable compounds.

Microalgae in the Biotransformation of Organic Compounds

Various microorganisms, mostly bacteria, fungi or yeasts, have been widely used in a range of biotransformation processes leading to the production of biologoically active compounds, but there is now growing evidence that some microalgae can also be effective in such processes. The increasing use in recent years of microalgae as agents in the biological transformation of organic compounds arising from a range of industrial processes has been reviewed by Ghasemi et al. (2011).

The advantage of using microalgae rests on the ease and relative low cost of their mass production. A range of microalgae, including the cyanobacterium *Anabaena cylindrica* and the green microalgae *Scenedesmus quadricauda* and *Coelastrum proboscideum* have been shown to be capable of transformating exogenously supplied steroids; a good example being the conversion of 4-androsten-3,17-dione to the equivalent hydroxylated derivative (Abul-Haij & Qian 1986). Microalgal transformation of progesterone, prednisolone and some other steroids has also been reported (Pollio et al. 1994, 1996). Studies with *Synechococcus* have shown that some strains of this cyanobacterium are able to transform several aryl methyl ketones to the corresponding (S)-alcohols, while others can transform aniline, naphthalene and phenol (Nakamura et al. 2000; Wurster et al. 2003).

The ability of isolated and purified cultures of *Synechococcus nidulans* to carry out the bioconversion of exogenously supplied hydrocortisone, one of the most useful intermediates for the production of a range of pharmaceutical products, was reported by Rasoul-Amini et al. (2010). Among the reactions characterised were those involved in cleavage of the steroid side chain, an important step in the production of high value therapeutic pharmaceuticals. The enzyme involved was described as being similar to the enzyme catalysing the conversion of cholesterol to pregnenolone in mammals.

Biotansformation of monoterpenes, the main constituent of many essential oils, has been shown to be accomplished in experimental systems where microalgae are immobilized in calcium alginate beads and supplied with a range of different substrates (Rasoul-Amini et al. 2011). The results confirmed that two species in particular, the green microalga *Oocystis* sp. and the cyanobacterium *Synechococcus* sp., yielded biotransformation products of considerable potential commercial value in the flavour and fragrance industry or as natural insecticides and antimicrobial agents. The immobilisation technique provides greater regulation of the metabolic processes and hence product formation. It increases the retention time, reduces the risk of washout and may also allow recovery and reuse of the microalgal cells resulting in simpler and usually cheaper processing compared with that requiring extraction and purification of enzymes (Hamada et al. 2003).

Chapter VII

Microalgal Toxins

Phytoplankton "blooms" occur when the environmental conditions allow certain microalgal species to pass through a period of rapid growth resulting in a build-up of the biomass of a single or a restricted number of species. Such blooms often affect the other biota in a variety of ways, most noticeably, and certainly best known, when the bloom comprises toxic species, that is, species that contain compounds (secondary metabolites) harmful to other organisms. Some of the most-studied of such toxic blooms are the so-called red tides of dinoflagellate species that result in concentration of the toxin through the food chain leading, in extreme cases, to such serious conditions as Paralytic Shellfish Poisoning (PSP) in humans. Other microalgal-induced harmful effects on humans include DSP (diarrhetic shellfish poisoning), SAP (amnesic shellfish poisoning) and NSP (neurotoxic shellfish poisoning) (Granéli & Turner 2006). Marine shellfish and fish farming projects are particularly affected by such red-tide events which can cause serious stock mortalities and financial losses (Smayda 1991). Of the known 5,000 or so microalgal species, no more than about 80 species are known to cause serious health effects in humans (Hallegraeff, 2003).

The range of toxic cyclic polyethers produced by some marine microalgae includes okadaic acid (C_{44} H_{68} O_{13}), dinophysistoxin and the neurotoxin brevetoxin (Garson 2001). Their severe toxicity is due in part to their lipophilicity conferred by their unique ladder-like molecular skeleton. Dinophysistoxin has been shown to have an acetate origin and to undergo carbon skeleton rearrangement resulting in the loss of several carbon atoms derived from the carboxyl group of acetate. Amphidinolides are also derived from acetate, again by way of the pathway of single carbon atom deletion.

The presence of toxins in some microalgal species and their absence from others is yet to be explained as also is the fact that some species contain both toxic and non-toxic strains. If the toxins served a protective role (e.g. against grazers), they might be expected to be present in all strains and in all growth stages, but this is not the case. Some of the major factors influencing toxin content in a range of microalgae have been summarised by Granéli & Flynn (2006). In general terms, toxin content is markedly affected by the nutrient conditions, with phosphorus limitation often enhancing cellular toxin content. In most instances, toxin content is lowest under nutrient-balanced conditions (i.e. N:P approx. 16:1) and highest under conditions of physiological stress. The complexity of the relationship between toxin production and various aspects of microalgal metabolism, and its variability among and between species and strains, makes it difficult to assign a clear function for these compounds in, for example, predator-prey interactions. Species known to produce harmful toxins can be found in various microalgal groups including dinoflagellates, diatoms, cyanobacteria, raphidophytes, pelagophytes, haptophytes, chrysophytes and prymnesiophytes (Granéli & Turner, 2006). Some species of the globally distributed diatom *Pseudo-ntizschia* are known to produce domoic acid (DA), the neurotoxin that causes amnesic shellfish poisoning (Lelong et al. 2012). DA production has been shown to be triggered by silicate and phosphate limitation as well as by low iron concentrations or high concentrations of copper.

Apart from their toxic effects on man and other animals, the range of metabolites produced by some microalgal and cyanobacterial species are of great clinical interest for their potential as possible anti-tumor agents. Study of the biosynthesis of such compounds could facilitate production of structural analogues with pharmaceutical significance. For most microalgal species, mass culture techniques have been devised to produce the source material under carefully controlled conditions and in the desired quantities. For some others, such as the dinoflagellates, their comparatively lower growth rates and more fastidious growth requirements, present some difficulties. These have been addressed by Beuzenberg et al. (2011) who devised an autonomously controlled photobioreactor for the enhanced production of toxins by three dinoflagellate species. They found that maximal bioreactor production of the toxin gymnodimine was considerably higher than could be achieved in bulk carboy culture and higher again when operating in contiuous mode. It was concluded that further scaling up to large-volume automated photobioreactors (up to 52L) could produce enough toxin to allow structural characterisation, analytical standard production and follow-up pharmaceutical appraisal.

Toxic Dinoflagellates

Paralytic and Diarrhetic Shellfish Poisoning

Certain species of dinoflagellates are responsible for some of the more dramatic of microalgal toxic effects that impact directly on human health. Of particular concern are those species implicated in the occurrence of Paralytic Shellfish Poisoning (PSP). The severity of PSP is a function of the concentrating effect of the original dinoflagellate toxin with progression through the food chain. Such concentrating effects rely on the tolerance to PSP toxins shown by species lower down the food chain (such as filter-feeding oysters, clams and mussels) resulting in vectoring of lethal doses to the higher trophic levels. The link between PSP and blooms of different dinoflagellate species was known well before the toxin itself had been identified.

The first of the PSP toxins to be isolated (from the Alaskan butter clam) was the neurotoxin saxitoxin (STX) (Schantz et al. 1957). Its chemical characterisation was later confirmed as also was its dinoflagellate origin as being from the marine species *Alexandrium catenella* (= *Gonyaulax catenella*) (Schantz et al. 1966). A number of structural derivatives of STX have since been described from a number of marine dinoflagellate species (Oshima et al. 1987). Particularly high PSP toxin levels were recorded in Norwegian and Danish mussels where the original source was identified as the dinoflagellate *Alexandrium tamarense* (*A. excavatum*), a species that is also responsible for direct or vectoral harmful effects on fish and crustacean larvae (Gosselin et al. 1989; Robineau et al. 1991). It should be noted, however, that toxicity of *Alexandrium tamarense* to some target organisms is not always directly proportional to the PSP content of the dinoflagellate. This has led to the suggestion (Flores et al. 2012) that other factors may be involved, including an associated production of harmful reactive oxygen species (ROS) and consequent ROS-induced lipid peroxidation. The PSP content of toxic dinoflagellate cultures is affected by a number of factors such as light intensity, nutrient availability and temperature (Anderson et al. 1990; Lim et al. 2006) and it has been suggested that the presence or absence of marine bacteria may also have an effect (Ho et al. 2006). Maas et al. (2010) incubated cultures of the toxin-producing dinoflagellate *Alexandrium minutum* under different conditiond and showed that saxitoxin was produced only in cultures exposed to natural daylight cycles, whereas gonyautoxin production was comparable in both dark- and light-grown cultures. Diarrhetic shellfish poisoning (DSP) has been well documented from the Mediterranean area and related to the presence in the water of various dinoflagellate species belonging to the genus *Dinophysis* (Sidari et al. 1998). Spring outbreaks of *D. sacculus* in a brackish lagoon in Sicily, Italy were associated with the production of okadaic acid and other shellfish poisoning toxins (Giacobbe et al. 2000). The principal toxins responsible for DSP belong to the okadaic acid group and its analogues (Quilliam 2003).

Brevetoxins and Haemolytic Effects

Of the dinoflagellate species now known to produce toxic compounds, a number belong to the family Kareniaceae. *Karenia brevis* (= *Gymnodinium* cf. *breve*) for example, produces powerful cyclic polyether neurotoxins, known as brevetoxins, which can cause extensive fish kills and massive marine mammal mortalities. Blooms of *K. brevis* have been reported to release the toxins into coastal waters where, under suitable conditions, they can become incorporated into aerosols, resulting in widespread respiratory illnesses in humans (reviewed by Kirkpatrick et al. 2004). More severe cases of neurotoxic poisoning are often compounded by bioaccumulation of the toxin in shellfish or other organisms in the food chain (Naar et al. 2004). Of particular concern are cases of haemolytic activity, affecting the pearmeability of red blood-cell membranes and often resulting in complete breakdown of the cells.

The toxic effects of *K. brevis* are shared by a number of other *Karenia* species. *K. selliformis,* for example, is the principal producer of toxic gymnodimines, spirocyclic imine compounds that have been shown to exert their major effects by targeting muscular and neuronal nicotinic acetylcholine receptors (Kharrat et al. 2008), as revealed in mouse bioassays. Other *Karenia* species have also been linked to extensive fish kills or strong haemolytic activity as well as observed mortalities of such diverse organisms as sea urchins, sea stars, snails, limpets, octopi and a range of fish species (Clément et al. 2001, Uribe & Ruiz 2001). Aspects of the biology and toxicology of different clones of *Karenia* were studied by Tatters et al. (2010) with a view to comparing the haemolytic activity of cells at different stages of a batch culture. The results were expressed as the cell numbers required to produce EC_{50} values in the standard red blood cell haemolysis bioassay. It was shown that both the cells and the cell-free medium of *K. selliformis* cultures produced significant levels of haemolysis over the entire period of culture. For other clones tested, haemolytic activity was detectable only later in the growth period and then only at very low levels compared with

K. selliformis. It is clear that haemolytic effects on a number of target organisms is a common feature across a wide range of dinoflagellate clones.

Ciguatera Fish Poisoning

Ciguatgera fish poisoning (CFP) is a very common form of human food poisoning in tropical and sub-tropical regions of the world. The symptoms, which include a variety of gastrointestinal, neurological and cardiovascular disturbances (see review by Palafox & Buenconsejo-Lum 2001) follow the consumption of fin fish contaminated by an assemblage of chemically related toxins (ciguatoxins) originating from a range of dinoflagellate species, especially belonging to the mostly benthic genera. Several species of *Prorocentrum* have been shown to produce the polyether toxin okadaic acid (e.g. Morton et al. 1998) commonly associated with diarrhetic shellfish poisoning in humans, and attributed to the consumptiom of contaminated shellfish. Similar health problems have been reported following the consumption of shellfish that have accumulated dinophysistoxin produced by planktonic dinoflagellates of the genus *Dinophysis* (Yasumoto et al. 1985).

The links between dinoflagellate population dynamics and the occurrence of various forms of ciguatera poisoning are not well characterised. The seemingly random occurrence of ciguatera events and the contradictory results obtained in different localities reflect our incomplete knowledge of the ecology of different toxic dinoflagellate species. In particular, many if not most of the ciguatera-related dinoflagellates are benthic forms (Anderson & Lobel 1987), necessitating consideration of assemblages rather than single bloom-forming species, and evaluation of the range of biotic and abiotic factors influencing passage of the toxin into the food chain. In an attempt to address this problem, Richlen & Lobel (2011) investigated the abundance of toxigenic benthic dinoflagellates of the genera *Gambierdiscus*, *Prorocentrum, Ostreopsis* and *Amphidinium* at Johnston Atoll, Pacific Ocean. They showed that both the total abundance of these toxic dinoflagellates and the community composition were different in different habitats and that the dinoflagellate community structure was primarily determined by the degree of water motion. This is consistent with observations (e.g. Ruff 1989) that distrurbance and destruction of coral reefs often precipitate ciguatera outbreaks.

The Toxic Pfiesteria Complex

The dinoflagellate genus *Pfiesteria* has attracted particular attentions because it includes a number of toxic strains with the potential to impact strongly on a range of fish species. In their review of the *Pfiesteria* complex of species/strains, Burkholder et al. (2001) describe the strong attraction that toxic strains have for live fish, causing stress, disease and, in extreme cases, death. Attempts to characterise the toxic effects of certain strains of *Pfiesteria* are made more difficult by their complex life cycles, with multiple amoeboid, flagellated and cyst stages, only some of which are toxic to humans or fish. They also have a cosmopolitan distribution and an ability to survive and grow under a wide range of conditions of temperature and salinity. Many of the strains loose their toxicity following artificial culture over a period of weeks to months and loose their capacity for induction to the toxic state. The

Pfiesteria toxin is water soluble and has its primary effect in the disruption of calcium metabolism of the affected fish or other prey. Aerosols from estuarine fish kills associated with this dinoflagellate have been reported to have respiratory and neurological impacts on human subjects.

Toxic Cyanobacteria

Cyanobacteria commonly produce a wide range of secondary metabolites of varying chemical type and abundance, many of which can cause acute and chronic poisoning in animals and humans. These problems are compounded by the ability of certain cyanobacteria (including the toxic species) to dominate the microalgal population of water impoundments to form blooms, scums and, in extreme cases, extensive mats. This they are able to do because of their tolerance of high temperatures and low incident light and their ability to vary their position in the water column by buoyancy control mechanisms. Other contributing factors would be an ability to fix atmospheric nitrogen and to store phosphorus within their cells. They are thus able to grow well in a variety of habitats such as estuaries, coastal lagoons, bays and water supply impoundments. Cyanobacterial blooms usually occur at higher temperatures and under conditions conducive to thermal stratification of the water column. The toxic species *Nodularia spumigena*, for example, forms extensive blooms in the Baltic Sea, with highest abundances at depths of 5m but often extending to depths of 18m due to its ability to acclimatise to different irradiance levels by adjusting the balance of photosynthetic pigments to maximise quantum efficiency at low light levels and to provide protection against the harmful effects of excessive light intensities (Jodlowska & Latala 2010).

Among the best known cyanotoxins are various cyclic peptides such as the hepatotoxic, tumour-promoting microcystins (MCs) and cylindrospermopsin as well as a range of alkaloids such as the potent neurotoxins anatoxin-a and saxitoxin (Chorus & Bartram 1999). Toxins produced by cyanobacteria are generally of amino acid or polyketide origin. They include a series of dolastatin compounds which are linear peptides (Garson 2001). Other novel nitrogenous toxic metabolites produced by some cyanobacterial strains, either as natural populations or when grown in culture, are the malyngamides debromoaplysiatoxin, lyngbyatoxins, curacins and barbamides. Biosynthesis of the last named has been studied using stable isotope incorporation experiments. It has been shown to involve chlorination of the pro-S methyl group of leucine, with chain extension by malonyl-CoA. The thiazole ring is thought to have a cysteine origin. Some cyanobacteria, in common with Gram-negative bacteria generally, also produce endotoxins (lipopolysaccharides, LPS) which are characteristic structural components of the outer cell membranes (Weckesser & Drews 1979). LPSs have been shown to affect the mammalian immune system leading to tumour necrosis. They may also interfere with liver functioning (Choi & Kim 1998). The most commonly used bioassay for cyanotoxins is the mouse bioassay, although such assays are now often replaced by bioassays based on the use of the brine shrimp *Artemia salina* (Metcalf et al. 2002) or species of *Daphnia* (Nizan et al. 1986).

The most frequently cited hypotheses to ascribe possible biological functions to cyanobacterial toxins relate to allelopathic effects such as feeding deterrence and inhibitory effects on other competing microalgae. Extracts of the toxic filamentous cyanobacterium

Lyngbya majuscula have been shown to deter feeding in a number of reef fish species and some other reef organisms (see review by Osborne et al. 2001) Large cyanobacterial mats produced under suitable conditions, although neither calcified nor otherwise physically resistant to herbivory, are probably protected from predation by their contained toxins. A number of cyanobacterial species have been implicated in large-scale fish kills with poisonous effects of their contained cyanotoxins being strongly implicated.

Microcystins

There are more than 80 microcystins, many of them known to cause waterborne illnesses and, in some instances, deaths of humans and animals. Of all the cyanobacterial species known to produce toxins harmful to man and other animals, perhaps the most studied is the freshwater species *Microcystis aeruginosa*. *Microcystis*, a member of the Chroococcales, occurs as spherical or subspherical cells which may be either solitary or grouped together in a mucilaginous matrix to form colonies, sometimes large enough to be visible in surface waters of lakes or slow-flowing rivers. *M. aeruginosa* produces the potent hepatotoxin microcystin LR and a range of other microcystins of similar toxic potency to microcystin LR (e.g. see Harada et al. 1990). Benthic cyanobacteria growing on sediments in reservoirs or on rocky surfaces in littoral or running waters are also a significant source of microcystins (Izaguirre et al. 2007). Kurmayer (2011) studied the production of microcystin (MC) and NP (nostophycin) in *Nostoc* sp. 152 under various conditions of temperature, irradiance and nutrient supply and showed that both MC and NP were constitutively produced, with the cellular content being significantly higher under physiological stress conditions. There was a linear correlation between the growth rate and both MC and NP production.

The biosynthesis of these cyclic peptides has been shown to involve a mixed acetate/amino acid pathway with additional single carbon atoms provided by methionine. Other commonly produced cyanopeptides, including aeruginosins, anabaenopeptides, cyanopeptolins, microginins, microviridins and oscillapeptins may also contribute to the bioactivity of cyanobacterial blooms (Rohrlack et al. 2005). The cyanostatins are linear lipopeptides and microginin analogues, most of them pentapeptides containing a characteristic novel amino acid, 3-amino-2-hydroxydecanoic acid. The biological activities of the cyanopeptides are wide ranging and include protein phosphatase inhibition and serine protease inhibition. The likely pharmaceutical potential of the cyanostatins is indicated by their high potency as inhibitors of leucine aminopeptidase (Sano et al. 2005). Codd et al. (2009) developed methods for the extraction and analysis of two cyanobacterial peptides, cyanostatins A and B, and were able to show that although microcystins are clearly the most commonly found cyanotoxins, some of the non-microcystin cyanopeptides often occur at much higher concentrations with potentially correspondingly high bioactivity.

Falconer (2005) has reviewed the extensive literature on the acute, chronic and subchronic toxic effects of microcystins on a range of organisms and has detailed the health effects on livestock and humans. The toxic effects appear to be largely centred on liver injury and hepatocyte degeneration, linked through specific inhibition of particular phosphatase enzymes. Other effects include tumour promotion, an effect that microcystins appear to share with other tumour-promoters of microalgal origin like the diarrheic shellfish poisoning factor okadaic acid (Falconer & Buckley 1989). The particular sensitivity of cells in certain organs has been attributed to the presence or absence of microcystin transporters which are known to be present in the cell membranes of hepatocytes and those lining the intestine, but largely absent from the membranes of other cells (Falconer 2005).

Anatoxin

The genus *Anabaena* includes common bloom-forming cyanobacteria of world-wide distribution. It produces anatoxin-a, a natural organophosphate that acts as a non-competitive inhibitor of acetylcholinesterase (AChE) activity, preventing the enzyme from hydrolysing the neurotransmitter acetylcholine. Among the commonly observed symptoms is the characteristic salivation in vertebrates which occurs in doses close to LD_{50} levels (20-40μg kg^{-1}) (Matsunaga et al. 1989). In extreme cases, the acetylcholine effects on membrane receptors cause continuous muscle stimulation which may result in convolution, suffocation, brain damage and even death (Carmichael & Falconer 1993). Neurotoxicity, traditionally tested by means of mice bioassays, now tends to be tested by invitro biochemical assays such as AChE inhibition (originally designed to assay organophosphate herbicides).

Another cyanobacterium, *Nostoc,* also produces anatoxins, as well as a number of other bioactive compounds such as depsipeptides and cryptophycins which act as inhibitors of tubulin polymerisation (Golakoti et al. 1995). Another inhibitor from *Nostoc,* nostocarboline, functionally similar to anatoxin, inhibits acetylcholinesterase (Becher et al. 2009).

Cylindrospermopsin

Cylindrospermopsin (CYN) is a hepatotoxic alkaloid produced by the cyanobacterium *Cylindrospermopsis raciborskii* first recognised as the causal agent of a severe outbreak of gastroenteritis, mainly among children, on Palm Island in Queensland, Australia (Hawkins et al. 1985; Griffiths & Saker 2003). The toxin is a water soluble cyclic guanidinium alkaloid containing a unique tricyclic hydroxymethyl uracil (Ohtani et al. 1992). Other cyanobacterial genera known to produce this toxin are *Anabaena, Aphanizomenon, Raphidiopsis, Lyngbya* and *Umezakia* with the toxin being most prevalent in strains from temperate and tropical freshwaters. Berry et al. (2009) confirmed strong toxic effects following injection of toxin extracts into zebrafish (*Danio rerio*) embryos. Direct immersion of embryos in extracts caused increased mortality and consistent, developmental dysfunctions.

The most likely mode of action of CYN is inhibition of protein synthesis (as revealed by *in vitro* assay systems based on rabbit reticulocyte lysate translation and by cellular inhibition of protein synthesis in mouse hepatocytes). However, acute toxicity of CYN to liver cells is thought to be caused by biotransformation products rather than through a direct effect on protein synthesis (Runnegar et al. 1995). Covalent binding and breakage of DNA has also been suggested as a possible mode of action (Shen et al. 2002).).

In view of the well-documented incidents of both human and livestock poisoning attributable to CYN (Falconer 2005), much research has been directed towards determining the minimum oral dose giving an adverse effect in mice bioassays and the maximum dose giving no adverse effect, the latter being particularly important for determining guideline values for safe drinking water (Humpage & Falconer 2003). The mechanism of CYN toxicity has been extensively studied using bioassays based on isolated and cultured rat hepatocytes (Runnegar et al. 1995). Since the toxin primarily affects protein synthesis, it is clear that a wide range of cells and tissue types will be susceptible. The presence within the CYN molecule of the nucleic acid base uracil has suggested the possibility for interaction with adenine groups in RNA and DNA and consequent carcinogenic effects (Falconer 2005).

Microalgae and Taste, Odour and Other Nuisance Effects on Water Supplies

Many of the unpleasant taste and odour episodes experienced in water supplies are known to be due to chemicals of microalgal origin. One such chemical is geosmin (trans-1, 10-dimethyl-trans-2-decalol) (Naes 1988), which is produced by the cyanobacterial species *Oscillatoria brevis* (Naes 1985) and by a number of other phytoplankton (Jones & Korth 1995). Geosmin is responsible for the characteristic earthy, musty taste and odour of the affected waters. The pathway of geosmin production in these microalgae is thought to be linked to isoprenoid synthesis leading to chlorophyll as is suggested by the inversely proportional relationship between chl *a* and geosmin under different conditions (Saadoun et al. 2001). At higher light intensities, for example, there was less chlorophyll *a* and a correspondingly greater synthesis of geosmin by *Anabaena* sp., while conditions that stimulated chl.*a* production, reduced geosmin production (Blevins et al. 1995). Off-flavour episodes in reservoirs, lakes and fishponds in China have been attributed to *Lyngbya* with similar inverse relationships between geosmin and chl *a* production (Zhang et al. 2009). Of particular relevance to water quality management was the observation that extracellular geosmin, as a proportion of the total, was highest during the stationary phase of growth at higher temperatures and optimum light intensity. Expression of a geosmin synthase gene in *Anabaena circinalis* under different conditions has been shown to be constitutive (Giglio et al. 2011).

Another terpenoid compound also of cyanobacterial origin, and known to be responsible for earthy or musty "off-flavours" is 2-methylisoborneol (MIB) (Martin et al. 1991). Like geosmin, MIB is particularly troublesome in the aquaculture industry. In the southeastern United States, for example, pond-raised channel catfish (*Ictalurus punctatus*) are known to acquire earthy and musty taints through the absorbtion of these compounds, with the cyanobacterium *Oscillatoria perorata* as the major causal factor (Schrader et al. 2004). Attempts to manage this problem include dosing the ponds with copper-based products (Schrader et al. 2005) and with synthetic herbicides such as diuron (3-[3,4-dichlorophenyl]-1,1-dimethyl urea) or, more successfully, with the commercial biocide SeaKleen, a quinone based natural product which is selectively toxic to blooms of *O. perorata* (Schrader et al. 2004), without seriously affecting the green microalgae and diatoms. Another MIB-producing cyanobacterium shown to be effectively controlled by the quinone anthraquinone-59 is *Planktothrix perornata,* also known to cause off-flavours in channel catfish aquaculture (Schrader 2007). Alternative, more environmentally acceptable treatments to reduce cyanobacterial related off-flavours in cultured fish species, involve the use of hydrophobic solids and liquids (e.g. polystyrene or paraffin) capable of absorbing the offending terpenoids (Kelly et al. 2006).

Excessive microalgal growths are usually the result of eutrophication caused by agricultural run-off and sewage discharges. They cause problems by impeding flow through drainage or irrigation channels and by blocking pumps and filters in water supply systems. The most effective control methods rely largely on the use of herbicides and copper treatments but such methods arc being increasingly questioned for their lack of selectivity and for possible long-term harmful environmental effects. Among the alternatives that have been attempted is the use of barley straw, whose effectiveness as an inhibitor of microalgal growth is said to be due to its stimulation of bacterial growth which, in turn, immobilizes dissolved phosphate, rendering it unavailable for microalgal growth (Wingfield et al. 1985). Welch et al. (1990) tested the beneficial effects of placing rotting barley straw at certain locations along a disused water-course (Chesterfield Canal, UK) and noted reduced growth of the filamentous microalga *Cladophora glomerata* although there was no noticeable effect on the low standing crop of phytoplankton. A follow-up laboratory study (Gibson et al. 1990) showed that rotting barley straw could inhibit the growth of several planktonic and filamentous microalgae and that the inhibitory effect was progressive during decomposition of the straw, reaching a maximum after six months. Significantly, there were no inhibitory effects from autoclaved straw and liquor from rotting straw was effective in inhibiting microalgal growth even after passage through a 0.2μm filter.

Many phytoplankton species (and aquatic bacteria) excrete high-molecular weight, carbon-rich compounds, collectively labelled as extracellular polymeric substances (EPSs). They represent an important component of the recycling of carbon in aquatic ecosystems and contribute significantly to microbial food web structures and carbon flux (Passow 2000). Among the phytoplankton, diatoms have been identified as major producers of EPS s especially when they occur in bloom proportions (Mari & Kiørboe 1996). Mass occurrence of marine mucilage caused by blooms of the diatom *Coscinodiscus granii*, were reported for Ariake Sound, Japan, resulting in major disruption of the fisheries industry (Fukao et al. 2009). In this case, as well as in others caused by blooms of other microalgal species (e.g. Claquin et al. 2008) the production of extracellular mucilage is strongly influenced by the growth conditions (Fukao et al. 2012).

Chapter VIII

Microalgae as a Renewable Source of Biofuel

The attraction of microalgae as a potential source of renewable and carbon-neutral fuel rests on their capacity, under appropriate conditions, for high oil productivity, equivalent to or exceeding that from oilseed crops, without compromising more traditional feed or food production. Microalgae, because of their wide physiological diversity, are able to grow in a range of marine and freshwater environments including hypersaline, brackish and highly alkaline waters. Such flexibility also creates the potential to allow mass culture with specific tailoring of the growth medium and conditions to increase the yield of the required product and to decrease contamination by other microorganisms.

For most microalgae, the default mode for photosynthetic carbon partitioning under favourable environmental or cultural conditions is 30-50% protein, 20-40% carbohydrate and 8-14% lipids on a total organic carbon basis. A large proportion of the lipids, 80% or more, is in the form of polar lipids (phospholipids and glycolipids) with only a minor component in the forms of neutral lipids (mostly triacylglycerol). With appropriate treatment, however, microalgal fatty acids can yield a mixture of monoalkyl esters suitable as a source of biodiesel fuel (Chisti 2007; Hu et al. 2008).

The advantages and disadvantages of biodiesel as a fuel source, compared with mineral diesel, have been documented by, among others, Hill et al. (2006) and Feofilova et al. (2010). Among the advantages are its renewable nature, its higher combustible value (and hence greater safety), its more benign impact on the environment (due to its ease of decomposition), its "cleaner" exhaust gases, its more complete combustion and its lower content of sulphur. Biodiesel from microalgae releases less carbon monoxide, hydrocarbons and sulphur oxides than diesel from fossil sources. Among the disadvantages are its higher viscosity (affecting the efficiency of fuel injection), its resistance to oxidation (correlated with the saturation level of the carbohydrate chains) and its lower crystalisation temperature. Biodiesel fuels from microalgal sources compared, for example, with those from plant sources, generally have a high level of unsaturation of their constituent fatty acids (Harwood & Guschina 2009), rendering them suitable for the production of biodiesel only after hydrogenation.

The first step in the production of fuel from biomass (pyrolysis) has been described as occurring more favourably from microalgal sources than from other sources, such as ligno-celluloses) (Peng & Wu 2000). It proceeds under lower temperatures and produces oil of higher quality. The most effective form of pyrolysis is the so-called "fast pyrolysis" which occurs in the absence of oxygen and under relatively low temperatures (450-550°C) and has been successfully used for the production of biofuel from *Chlorella protothecoides* and *Microcystis aeruginosa* (Miao & Wu 2004, 2006). Alternative methods of obtaining biofuel from microalgal biomass include thermomechanical liquefaction (which is carried out at 300°C and 10MPa pressure and uses sodium carbonate as catalyst) (Dote et al. 1994) and gasification (which produces a methane-enriched fuel) (Minowa et al. 1995). Methane can be produced from microalgal biomass by a process of anaerobic digestion. Zamalloa et al. (2012) subjected *Phaeodactylum tricornutm* biomass to a novel form of biomethanation in an anaerobic membrane bioreactor. By regulation of the loading rates, retention times, membrane fluxes and the bioreactor microbial communites, biogas recovery of the order of 52% of C input was obtained. In addition, the clear effluent from the bioreactor was rich in inorganic nutrients usable as liquid fertilizer.

The Biochemistry of Biodeisel Production by Microalgae

The main components of the microalgal source material for biodiesel production are triglycerides, which may, under suitable conditions, constitute a significant proportion of the total biomass. This lipid material is first transesterified by treatment of the triglycerides with alcohol (most commonly methanol or ethanol) in the presence of a catalyst to yield three molecules of fatty acids (e.g. fatty acid methyl esters, FAMEs) and one molecule of glycerine per triglyceride molecule:-

$$
\begin{array}{lcccll}
CH_2-O-CO-R_1 & & & & R_1-CO-OCH_3 & CH_2-OH \\
| & & & & & | \\
CH-O-CO-R_2 & + & 3CH_3OH & = & R_2-CO-OCH_3 & CH-OH \\
| & & & & & | \\
CH_2-O-CO-R_3 & & & & R_3-CO-OCH_3 & CH_2-OH \\
\text{Triglyceride} & & \text{Methanol} & & \text{Fatty acid Esters} & \text{Glycerol}
\end{array}
$$

The reaction occurs by way of a series of stages and under conditions carefully selected to suit different triglyceride sources. Catalysts, such as acids and alkalis, are also required to ensure a good yield of biodiesel, with alkaline catalysis reported as being much more rapid than acid catalysis (Fukuda et al. 2001). The by-product of diesel production from triglycerides, glycerine, is itself a valuable product with the potential to serve as the source molecule for a variety of useful compounds such as acetone, butanol or propionic acid (Lee et al. 2001). The use of enzymes (lipases) for transesterification, although costly, offers an environmentally friendly option as a realistic alternative to other methods and has been tested for its commercial viability. In the process flow system for transesterification of lipids devised by Xiong et al. (2008), lipase (triacylglycerol acylhydrolase, EC 3.1.1.3 from yeast) was used, with the enzyme immobilised by physical absorption on to macroporous resin.

The preferred lipids for biodiesel production are neutral lipids, mainly triacylglycerols (rather than phospholipids or glycolipids). Oleic acid is particularly favoured because of the greater stability of its alkyl esters and the better characteristics of the fuel produced (Knothe 2005). Enzymic reactions contributing to biodiesel production include the desaturases, which act by introducing a double bond at a specific position in the hydrocarbon chain. Acyl-ACP desaturases, such as the stearoyl-ACP desaturase (SAD) common in microalgae, converts stearic acid (18:0) into oleic acid (18:1) (Tocher et al. 1998).

Compared with vegetable oils, oils from microalgae are richer in polyunsaturated fatty acids (some with four or more double bonds). They are therefore prone to oxidation unless the degree of unsaturation is lowered by partial catalytic hydrogenation. In its main characteristics, however, biodiesel from microalgae is quite similar to petroleum diesel (Amin 2009). The different carbon chain lengths and the number and positioning of unsaturated bonds all influence the CN number of the fuel, that is, the measure of its tendency to ignite in the engine. Other features of the fuel influenced by the fatty acid characteristics are its oxidative stability, cold-flow properties and viscosity (Ramos et al. 2009). Oils rich in linoleic and linolenic acids tend to have poor oxidative stability and most biodiesel from such sources requires the addition of antioxidants. Saturated fatty acids, particularly those with long chain lengths, confer a relatively high melting point and it has been suggested that best results are obtained with a mixture of saturated and unsaturated fatty acids (Knothe 2009).

Prefered Microalgal Species for Biofuel Production

Selection of suitable microalgal species for biofuel production requires careful evaluation of a range of different characteristics including the rate of biomass production and its lipid characteristics, tolerance of environmental conditions and resistance to contamination, ease of harvesting and subsequent processing. Griffiths et al. (2011) compared the lipid productivity and fatty acid profile of 11 microalgal species cultured in an airlift photobioreactor. With a view to minimising harvesting costs, they also investigated the settling characteristics. Lipid yields and productivity were, in general, higher under low nitrate conditions, except for the cyanobacterium *Spirulina platensis.* Highest growth rates, and greatest increase in lipid content in response to nitrogen limitation, were obtained with *Chlorella vulgaris* and *Scenedesmus* sp. Fastest settling rates, and highest biomass recovery after 24h gravity sedimentation, were obtained with *Cylindrotheca fusiformis, Scenedesmus* sp., *Tetraselmis suecica* and *Spirulina platensis.* The most promising species, in terms of meeting European standards for biomass production, were the freshwater microalgae *Scenedesmus* sp. and *Chlorella vulgaris* and the two marine species *Cylindrotheca fusiformis* and *Nannochloropsis* sp.

Lipid productivity can be defined as the product of lipid content and biomass productivity. It is clear from the studies that have been published (e.g. Griffiths & Harrison 2009) that lipid content is not generally a reliable indicator of lipid productivity. Lipid content values in N- or Si-deficient conditions were reported by them to be, on average, between 138 and 168% respectively of those reported under the equivalent sufficient conditions. In terms of lipid productivity, however, lipid content and growth retardation under the deficient conditions would tend to be counterbalanced. Selecting species for optimum biodiesel production would also require information on a range of other considerations such as resistance to contamination and tolerance of the operating conditions employed. Also important would be ease of harvesting and downstream processing.

Similar results were reported by Li et al. (2008) for the chlorophyte *Neochloris oleoabundans* with the highest cell lipid content (0.40 g/g) recorded in cultures grown at low N conitions (3 mM sodium nitrate) but best lipid productivity (0.133g.$L^{-1}d^{-1}$)was achieved at 5mM sodium nitrate (lipid content 0.34g/g and biomass productivity 0.40g.$L^{-1}d^{-1}$). Highest biomass productivity (0.63g$L^{-1}d^{-1}$) was obtained at 10mM sodium nitrate.

Triglycerides from Chlorophytes

Among the microalgae that have been tested for their suitability for biodiesel production are species belonging to the chlorophyte genera *Chlamydomonas, Chlorella, Nannochloris,* and *Nannochloropsis.* Diesel from *Chlorella* was found by Xiong et al. (2008) to have a higher heating value than that from vegetable oil (41Mj kg^{-1} v. 37.3 - 39.8 Mj kg^{-1}) and a lower density (0.864 kg l^{-1} v. 0.873 - 0.885 kg l^{-1}). Almeida et al. (2009) tested 31 cultured microalgal strains and selected 11 of the faster-growing strains for their yield of lipid and triacylglycerol (TAG). One species of *Nannochloropsis* (*N. gadiata*) gave the highest yield of lipid and TAG while another (*N. occulata*) also gave the high values for TAG productivity (up to 20mg TAG $L^{-1}day^{-1}$). Among the other chlorophytes known to accumulate large amounts of lipid material are *Muriellopsis* and *Pseudochoricystis* (Ito et al. 2009). In the latter, as is the case generally with many microalgal species, the yield of triglycerides was much higher following culture under nitrogen-deficient conditions. Another important consideration is the selection of strains suitable for the particular region. To this end, Talukdar et al. (2009) tested 22 freshwater microalgal species isolated from diverse habitats in Assam, India. The fastest growers were local strains of *Chlorella* (μ = 0.3087) and *Botryococcus braunii* (μ = 0.2418). Strains of *Haematococcus* sp. (μ = 0.0921) and *Scenedesmus dimorphus* (μ = 0.0678) grew much slower (μ = specific growth rate, i.e. first-order reaction rate constant). Total lipid content varied between strains with *B. braunii* considered as offering the best potential for biofuel production.

Large ponds (*ca.*1000 m^2), seeded with different chlorophyte species, have been shown to be capable of yielding more than 50g dry matter per m^2 per day. Replacement of such ponds by bioreactor chambers has been suggested as a means of over-coming limitations to productivity imposed by seasonal and diurnal fluctuations and to eliminate bacterial contamination. Another possible development is the linking of such bioreactors to waste heat sources from certain industrial operations. The ability of many chlorophyte species to assume a heterotrophic or mixotrophic form of carbon assimilation, where photosynthetic carbon fixation is supplemented by the utilization of organic sources of carbon, provides another means of substantially improving productivity (Ratledge & Wynn 2002). Gardner et al. (2010) compared growth and triacylglycerol (TAG) accumulation in two species of chlorophyceae, *Coelastrella* sp. (PC-3) and *Scenedesmus* sp. (WC-1), the former isolated from a near-neutral pH source and the latter from an alkaline stream (pH 8.7). Growth rates and TAG accumulation were monitored under different pH levels and the effect of nitrate depletion was also observed. Both microalgal species grew under all the conditions tested and

TAG accumulation was enhanced by both high pH and by nitrate depletion. *Scenedesmus* grew best and accumulated more TAG under alkaliphilic conditions (pH>9).

Lipid accumulation in cultures of *Scenedesmus obliquus* was greatest under conditions of nitrogen deficiency (up to 43% of dry weight, compared with 12.7% in control cultures) (Mandal & Mallick, 2009). Cultures under P-deficiency or under thiosulphate supplementation (to provide a reducing environmenty and hence an increased pool of NADH) showed an increase in lipid content to 30% of dry weight. Cells of *Scenedesmus* pre-grown in a medium supplemented with glucose accumulated lipid (with palmitate and oleate as major constitutents) to 2.16g L^{-1} i.e. 40-fold higher than in control cultures. A study of the molecular and cellular mechanisms of the increased TAG accumulation by *Chlamydomonas reinhardtii* under unfavourable growth conditions has suggested that the increase may be due to degradation of chloroplasts and other intracellular membrane systems catalyzed by phospholipid: diacylglycerol acyltransferase (Yoon et al. 2012).

The colonial green microalgae *Botryococcus* (Trebouxiophyceae) has long been known to synthesise significant quantities of hydrocarbons whose chemical nature varies between different strains and between different conditions of growth (Casadevall et al. 1984; Banerjee et al. 2002; Metzger & Largeau, 2005). Cultivated under aeration with ambient air enriched to 10% CO_2, *Botryococcus braunii* yielded 26.6 mg biomass L^{-1} d^{-1} of which 21% was lipid material (Oh et al. 2009). Aeration with fluegas yielded a 2.9-fold increase in biomass productivity and a 3.8-fold increase of lipid productivity, with 55% of the fatty acids in the form of oleic acid, a suitable lipid source for biofuel production. Of the various cell-disruption methods tested for extraction of oleic acid, the microwave oven method was the most effective. Some strains of *B. braunii* produce linear olefins, chiefly C_{27}, C_{29}, and C_{31} dienes (Gelpi et al. 1968; Knights et al. 1970) that can constitute up to 36% of dry weight. Others produce branched olefins (botryococcenes) (C_nH_{2n-10}; n = 30-37), in one example contributing 86% of the dry weight (Maxwell et al. 1968). A range of complex botryococcenes have now been identified and structurally characterised.

A strain of *B. braunii* was found by Wolf et al. (1985) to produce up to ten different botryococcene compounds at a total concentration varying from 25 to 40% of dry weight, depending on the culture conditions. Aeration of the cultures with an enriched CO_2:air mix (0.3%), in particular, improved both growth rates and yield of botryococcenes. Up to 7% of the botryococcene fraction was found by Wolf et al. (1985) to be intracellular and the remainder was located within the colonial matrix, the latter being largely composed of the longer chain forms (C_{33} and C_{34}). These observations were interpreted to indicate an extracellular location for at least the terminal steps of biosynthesis of the hydrocarbon.

Some strains of *Botryococcus* produce long straight-chain odd-carbon-numbered hydrocarbons (mono- to tetra-unsaturated) from C_{23} to C_{33} derived from fatty acids, typically alkadienes of the general formula C_nH_{2n-2}. Others produce branched triterpenoids from C_{30} to C_{37} of the formula C_nH_{2n-10} called botryococcenes while yet others produce acyclic tetra terpene hydrocarbons called lycopadiene, typically $C_{40}H_{78}$ (Metzger & Largeau 2005). The biosynthesis of fatty acids by *Botryococcus braunii* was shown by Choi et al. (2011) to be markedly affected by the conditions of culture, in particular by the availability of a suitable supply of nitrogen. While maximum biomass production was obtained in a nitrogen sufficient medium (3.66mM nitrate), maximum lipid content (63% of dry weight) was obtained under strongly nitrogen-limiting conditions (0.04mM nitrate). Maximum lipid productivity (0.019g L^{-1} day^{-1}) was achieved in cultures supplied with 0.37 mM nitrate. Similarly, while oleic acid (an important component in biodiesel production) accounted for 86% of the total fatty acids under N-limited conditions, the equivalent value under N-sufficient conditions was 69%. The enzyme Acyl-ACP desaturase (which catalyses the conversion of stearic acid [18:0] into oleic acid [18:1] by introducing the double bond at a specific position) was also affected. The expression of the gene encoding for this enzyme was found by Choi et al. (2011) to be 2.6-fold higher under N-limited conditions compared with N-sufficient conditions.

A survey of a wide range of *Botryococcus* strains has identified three different races, depending on the type of hydrocarbons produced. Race A produces C25 to C31, odd-numbered n-alkadienes and alkatrienes (Bannerjee et al. 2002). The B race strains produce methylated triterpenes (botryococcenes) $C_n H_{2n-10}$, where n=30-37 (Raja et al. 2008). Races A and B are widely distributed in brackish and freshwater lakes while another race (L) is found only in water samples collected in the tropics. In their comparison of six different strains (four from Race A and two from race B) Eroglu et al. (2011) observed that the B-race strains accumulated substantially higher levels of botryococcene hydrocarbons (20 30% of overall biomass) than did the A-race strains (10-14%).

Another species (*B. sudeticus*) had a very low level of hydrocarbon accumulation of only about 3% of overall biomass, pointing clearly to the need for careful screening in the selection of suitable strains for commercial exploitation. Extraction of oil products from microalgal biomass usually requires disruption of the cell walls which, in most species, are strongly resistant to disruption. Some atrains of *Botryococcus* provide the exception in excreting oil form their cells.

The high-lipid-producing chlorophyte *Tetraselmis suecica* (Prasinophyceae) can achieve an average lipid productivity of 32 mg L-1 day^{-1} (Griffiths & Harrison 2009), when grown in a photobioreactor. Under conditions not conducive to fast growth, a high proportion of the lipids present are neutral lipids. Using flow cytometry in conjunction with the lipid-soluble fluorescent dye Nile Red, Montero et al. (2011) were able to isolate and select cells with a higher and more stable lipid content from wild (and fast-growing) strains of this chlorophyte. They concluded that their sorting method, when allied to improved methods of securing entry of the dye into microalgal cells with thicker and more rigid cell walls, might be usefully applied in the search for other high-yielding oleaginous microalgal strains.

Under semi-continuous growth in 120L bag photobioreactors, Mohemani (2013) was able to grow *T. suecica* and *Chlorella sp.* with yields of 110 and 140 mg DW.L^{-1}d^{-1}respectively over an 11month period in Melbourne, Australia, with minimum lipid content of 25%. Interestingly, several attempts to cultivate *Dunaliella tertiolecta* under these conditions were unsuccessful.

Triacylglycerol from Diatoms

In the survey carried out by the US Department of Energy in the 1990s, diatoms were identified as strong candidates as potential feedstock for triacylglycerols (TAGs) and hence for biodiesel production (Sheehan et al. 1998).

Marine species, and those tolerant of saline or brackish conditions, were particularly favoured. It was further shown that TAG production by such species is highest under conditions of environmental stress such as nutrient deficiency or low temperature (see review by Hu et al. 2008). Yu et al. (2009) studied both the yield and the molecular characteristics of

TAG in two species of diatoms (*Thalassiosira pseudonana* and *Phaeodctylum tricornutum)* as a function of nitrate and silcate depletion. By using a combination of analytical techniques, they were able to show that, in batch cultures, TAG accumulation starts approximately 10h after the cultures have entered the stationary phase of growth and, under silica starvation, can reach *ca.*18% of total dry weight in the 3H strain of *T. pseudonana* and *ca.*14 % in *P. tricornutum*.

Silica starvation yielded more TAGs in *T. pseudonana* than did nitrate starvation. However, the TAG molecular species were similar for both diatoms tested, confirming previous reports that the predominant fatty acids in these species were of the 14:0, 16:0, 16:1, 16:3, and 20:5 form. Such low molecular weight polyunsaturated fatty acids would have good cold-flow properties but would be susceptible to oxidation, unlike saturated fatty acids of longer acyl chains which have higher energy yields and superior oxidative stability but a greater susceptibility to solidify at ambient temperature. Results such as these and the observation that TAG accumulation is a regulated process, opens up the possibility of future genetic manipulation to improve the yield and nature of biodiesel feedstock from such sources.

The acceptability of the chosen diatom source material for biodiesel production depends primarily on the amount and chemical nature of the lipid fraction. The world-wide distribution of the genus *Skeletonema*, and its prominence as a major component of the phytoplankton of coastal Chinese waters (Yan et al. 2011) has attracted particular attention to this diatom as a possible source of photosynthetic glycerolipids suitable for biofuel production.

Popovich et al. (2012) compared *Skeletonema costatum* and *Navicula gregaria* and were able to confirm that both species had a significant proportion of their total lipid content in the form of neutral lipids (65% in the former, 76% in the latter). In *S. costatum,* saturated fatty acids (SFAs) accounted for 43.48% of the total, with a similar proportion (40.11%) in the form of monounsaturated fatty acids (MUFAs) while in *N. gregaria*, the proportions were 33.42% for SFAs and 54.85% for MUFAs.

In terms of their mix of fatty acid types, both species were deemed to be suitable as source material for biodiesel production, with an acceptable cetane number (a measure of the fuel's ignition quality) albeit with less than ideal cold-flow properties. Popovich et al. (2012) conclude that the characteristics of the source material from these two species are such as to offer good potential for further refinements to improve biodiesel quantity and quality through improved production and harvesting methods, through blending with other products or through genetic modification of the microalgae.

Nannochloropsis (a member of the Eustigmatophyceae and hence grouped with the diatoms in the Heterokontophyta) is well known for its high content of neutral lipids (Sukenik,1999) which, as is the case with this group generally, are characterised by four particularly abundant fatty acids: palmitic (16:0), palmitoleic (16:1n-7), arachidonic (20:4n-6) and eicosapentaenoic acid (20:5n-3). The combined proportions of the 16:0 and 16:1 fatty acids can account for as much as 70-80% of the total fatty acids, especially under stress conditions.

Biofuel from Cyanobacteria

Given their observed capacity for efficient triacylglycerol (TAG) accumulation, it is not surprising that much of the current effort towards biofuel production from microalgal sources has been directed towards the eukaryotic microalgae. Cyanobacteria appear to produce little or no TAGs but it has been suggested (Jansson, 2012) that the cyanobacteria may nevertheless be amenable to metabolic engineering approaches to redirect carbon flux in lipid metabolism towards the synthesis of molecules more suitable for biofuel production. One advantage that cyanobacteria might offer is the fact that, unlike the situation in eukaryotic microalgae, all their metabolic processes are directed by soluble or membrane-bound enzymes in the cytoplasm.

Another is that their capacity for homologous recombination, which allows rapid site-directed mutagenesis, gene insertion, replacements and deletions in what Jansson (2012) has described as a precise, targeted and predictable manner.

It is anticipated that such genetic engineering methods might redirect cyanobacterial metabolism towards the synthesis of hydrocarbons such as alkanes and isoprenoids, as has already been achieved with heterotrophic bacteria (Schirmer et al. 2010). The metabolic flexibility of cyanobacteria is already evident from their demonstrated capacity to produce alcohol-based fuels such as ethanol and isobutanol (Deng & Coleman, 1999, Atsumi et al. 2009).

Schemes for linking cyanobacterial photosynthesis and hydrocarbon fuel production to point sources of CO_2 have been suggested (Jansson 2012) although a number of related practical problems would have to be solved. It has also been suggested that it might be possible to develop cyanobacterial strains that secrete alkanes or fatty acids which then float to the surface of the reactor vessel (Kaczmarzyk & Fulda 2010).

Maximising Lipid Production by Cultured Microalgae

Two-Stage Cultivation

A number of culture studies have shown that accumulation of triacylglycerols (TAGs) by microalgae is highest under growth-limiting conditions such as high salinity or nitrogen deficiency (Solovchenko et al. 2008, Takagi et al. 2006) but since these conditions sustain only low biomass yields, overall lipid poductiviy is low. In an attempt to overcome this problem Su et al. (2011) devised a two-stage culture process comprising separate phases of growth and lipid production.

During the growth phase the cultures are maintained for a period under nitrogen sufficient conditions (to build up biomass) and then transferred to nitrogen depleted conditions for the lipid accumulation phase. Results for cultures of *Nannochloropsis oculata* are shown in Figure 11, which compares biomass and lipid production in a single stage and a two-stage cultivation process. In a single-stage batch culture under nutrient sufficiency in the light the biomass increases over the first 5 days to a maximum of 0.58 g L^{-1} before entering into a 4-day stationary phase during which the biomass increased only slightly to 0.654g L^{-1} (see Figure 11a).

Over the first 5 days, the lipid content remains low and constant at ca. 10% but increases to 15.8 % during the subsequent 4-day stationary phase. For the two-stage culture method, the cells were harvested at the end of the first 5 days of growth, washed and transferred into a nitrogen deficient culture medium for the second stage during which the biomass increased to 0.86g L^{-1} (Figure 11b). But the lipid content increased to 48.2%. Under the two-stage strategy, lipid content increased three-fold and the average rate of lipid accumulation was 7.7 times higher than under the single stage batch cultivation.

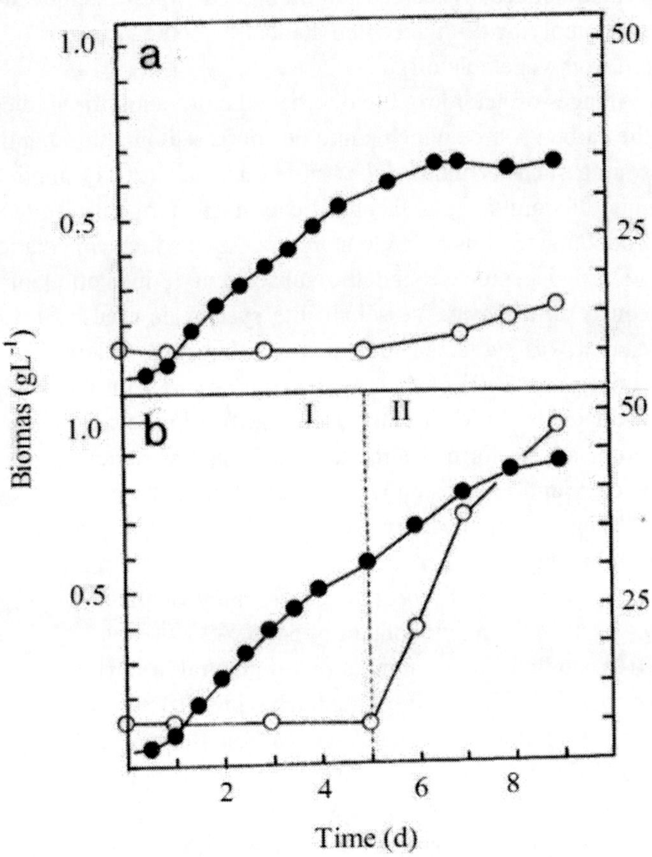

Figure 11. (a) Biomass (closed circles) and lipid (open circles) accumulation in a single-stage batch culture of *Nannochloropsis oculata* under nitrogen-sufficient conditions in the light (300 µmol photons m^{-2}) and with 2L min^{-1}aeration. (b) Corresponding biomass (closed circles) and lipid (open circles) data for cultures exposed to a two-stage regime for which the cells were harvested at the end of the first 5 days (vertical broken line) and transferred to nitrogen-deficient conditions (other conditions as in (a) but with aeration with 0.5 L min^{-1} 2% CO_2–enriched air). (Redrawn, with permission, from Su et al. 2011).

Culture under Heterotrophic Conditions

Many microalgae are able to supplement their autotrophic nutrition with a heterotrophic form of nutrition which uses an organic substrate as a source of energy and carbon. Based purely on energy considerations, photoautotrophic cultivation of microalgae, in open ponds or in photo-bioreactors using solar energy, would appear to be the more economical method. However, heterotrophic nutrition, by eliminating (or reducing) the dependence on light, is particularly suited to biomass production by fed-batch culture because it allows enhanced cell density and hence product concentration. Fed-batch culture minimizes substrate restriction by maintaining continuous feeding of the nutrient stream thus allowing prolonged exponential growth. Using the microalga *Chlorella protothecoides*, in an improved high-density fed-batch bioreactor with regulated substrate feeding, Xiong et al. (2008) achieved cell densities up to 51.2 g L^{-1} in 167 h in a 5L bioreactor and a lipid content of 50% or greater of cell dry weight. Subsequent transesterification catalyzed by immobilized lipase achieved up to 98% conversion to biodiesel complying with accepted standards (some superior to those achieved by biodiesel production from vegetable oil).

Among the disadvantages of heterotrophic microalgal cultivation for biodiesel production was the high cost of the carbon source which could be addressed only through the use of more economical carbon source such as industrial waste. Lu et al. (2011) applied to biodiesel production the technique of simultaneous saccharification and fermentation (SSF) previously used for bioethanol production. SSF is a single-step process in which enzymatic hydrolysis of the source material to produce glucose and the subsequent fermentation of the sugar are carried out simultaneously in the same vessel. In the system devised by Lu et al. Cassava (*Manihot esculenta*) starch was subjected to α-amylase digestion followed by heterotrophic cultivation of exponential phase cells of *Chlorella protothecoides* on the resulting substrate. In the SSF procedure adopted by Lu et al. (2011) saccharification (enzymic starch hydrolysis) and glucose fermentation are performed simultaneously in the same vessel. This provides improved energy efficiency and reduces apparatus and equipment costs. The authors claim a high conversion rate of sugar to biodiesel (66.67 % mole product C per mole sugar C) with the resulting biodiesel having a higher heating value than is obtained from bioethanol systems. The maximum biomass of *C. protothecoides* they obtained from a 5L fed-batch culture vessel was 49.43 g L^{-1} with an oil content of 54.60%.

Xiong et al. (2010) studied the accumulation of neutral triacylglycerols in *Chlorella protothecoides* grown in chemically defined media having different carbon-nitrogen (C/N) ratios. Using ^{13}C-tracer technology they analysed carbon flux through different pathways based on recognition of ^{13}C labelling patterns imprinted in different metabolic intermediates (e.g. protein-bound amino acids). In experiments in which C1-^{13}C labelled and Universally-^{13}C labelled glucose was supplied to support heterotrophic growth of *C. protothecoides,* gas chromatographic and mass spectrometric analysis revealed a metabolic network consisting of glycolysis, pentose phosphate pathway and the tricarboxylic acid cycle, but no evidence of an active glyoxylate shunt. There was also evidence of phosphoenol pyruvate (PEP) carboxylase, PEP kinase and malic enzyme activity.

When the cultures were grown under nitrogen-limited conditions, there was an increase in pentose phosphate activity relative to glycolysis pathway activity and a consequent greater generation of NADPH which, in contrast to NADH, is specifically utilized in the biosynthesis of long-chain fatty acids and hence in the biosynthesis of lipids. Thus, in media having high C/N ratios (i.e. N-limited conditions) 40.5% of the supplied glucose (as Glc-6-P) enters the pentose phosphate (PP) pathway and 53.3% was catabolised through the glycolysis pathway, with 81.0% of the glucose input flux directed to *de novo* synthesis of fatty acids. In media having low C/N ratios (N-sufficient conditions) carbon flux through the PP pathway accounted for only 33.3% of the glucose input flux and carbon flow into the pool of fatty acids (via acetyl-CoA) was 54.2% of the glucose input flux. These estimated net fluxes were percentages of the relative rates normalised for the rates of glucose uptake which were 0.80mmol.g dry wt.$^{-1}$h^{-1} in nitrogen sufficient cultures and 65mmol.g dry wt.$^{-1}$h^{-1} in nitrogen limited cultures. Using heterotrophically grown cultures of *Auxenochlorella protothecoides* supplied with glucose and glycine as C and N sources respectively, De la Hoz et al. (2010) were able to select conditions which maximized both biomass (80gL^{-1}) and biomass productivity (19gL^{-1}d^{-1}) while still preserving a cellular oil content of over 50%.

Cultured cells of the colonial chlorophyte *Botryococcus braunii* produce large amounts of heavy oils (liquid hydrocarbons) when their photosynthetic productivity is supplemented by an additional heterotrophic or mixotrophic nutrition (Weetall 1985; Tenaud et al. 1989, Banerjee et al. 2002). Tanoi et al. (2010) investigated the effect of various organic carbon sources on the growth and morphology of an axenic strain of *B. braunii* in both light and darkness. They found that with strain B70, isolated from a pond in Higasi-Hiroshima, Japan, growth under a 12hL:12hD regime was greatly enhanced by the addition of various sugars, but especially glucose and mannose at concentrations greater than 10mM (maximum specific growth rates 0.15 – 0.16 day $^{-1}$). Interestingly, the size of oil granules in cells of *B. braunii* was greatly increased in the presence of glucose suggesting that addition of a carbon source may be a way of enhancing oil production in slow-growing species.

Reports that some cyanobacterial strains can also have their growth enhanced under mixotrophic conditions point to the possibility of improved yield under photobioreactor conditions where light would normally be a limiting factor. Yan et al. (2012) studied the energetics of the growth of a strain of *Synechococcus* supplied with glucose or acetate compared with photoautotrophic conditions. Energy conversion efficiencies, expressed as ATP synthesis under photoautotrophic, glucose-mixotrophic and acetate-mixotrophic culture were 4.59%, 5.86% and 6.60% respectively.

Economic Considerations

Borowitzka (1999) valued mass-cultured microalgal biomass then in use for commercial production of high value compounds to be in excess of \$10 (US).kg^{-1} and for profitable biofuel production, the equivalent cost would have to be no more than \$1.0 (US).Kg^{-1}. This ambitious goal, he goes on to predict, could be contemplated only under the most favourable conditions of year-round microalgal growth, harvesting and downstream processing, using the most efficient strains. A further requirement would be to utilize areas (e.g those where the only available water is saline) not suitable for crop production. It would be necessary therefore to select microalgal strains able to grow and maintain high lipid production at or above sea-water salinities (approx. 3% NaCl or above). Moheimani & Borowitzka (2006) report that the coccolithophorid *Pleurochrysis carterae,* when grown in outdoor raceway ponds of high salinity could maintain an everage annual productivity of >20g dry wt.m^{-2}day^{-1}.

In assessing the practicality of biodiesel production from microalgae, it is instructive to draw comparisons with another well established bioenergy source namely bioethanol production from sugar cane. It has been estimated that in terms of energy efficiency microalgal biodoesel is far superior to bioethanol (Gray 2006). When consideration is also given to the importance of cane as a food source, and to its requirement for prime agricultural land, the case for microalgal biodiesel becomes even stronger. A further possible factor in favour of a microalgal-based energy production system is the greater scope for increased productivity from genetic improvements and innovations of metabolic engineering. As an alternative to using microalgal biomass to produce biodiesel there have also been trials in which it is used to produce bioethanol. This involves subjecting the microalgal biomass after preliminary treatment, to a process of fermentation using yeast (*Saccharomyces cerevisiae*). Biomass from the green microalga *Chlorococcum* has been used in the production of bioethanol with yields considered to be commercially feasible when the product is used as a pure fuel or as an addative to other fuels (Harun et al. 2010).

The U.S. Department of Energy Biomass Program, in its draft report (2009) estimated that had the 67 million acres of soybeans cultivated in 2007 been given over entirely to biodiesel production, it would have satisfied only 6% of the country's on-road petroleum diesel use; the same acerage devoted to microalgal culture has the potential, they claim, to satisfy more than 100% of annual diesel usage. The Oil Market Report of the International Energy Agency (cited in Wijffels & Barbosa 2010) estimated that replacing all of Europe's transport fuels by biodiesel from microalgal sources would require 9.25 million ha under culture, assuming a productivity of 40,000 litres per ha per year. This assumption is based on a 3% solar energy conversion to biomass with an oil content of 50%. The enormous scaling-up that would be required to get anywhere near these targets for commercial microalgal biodiesel production could only be contemplated if supported by significant technological advances, including strain improvements.

Estimations of the cost of producing microalgal biomass from photobioreactors or from open cultivators have been placed at approximately $1-10(US).kg^{-1} (Chisti 2007). Assuming a 30% yield of oil per unit microalgal biomass, this has been equated to a biodiesel cost of $3-30 (US) per litre. Varfolomeev & Wasserman (2011) give a figure of $2.8 (US) per litre. For such yields to be competitive would require extensive reduction in the costs of production and/or great improvement in production efficiency. The assessment by Pimentel (2008) of the practicality of oil production from microalgae concludes that even with the most optimistic productivity assumptions, the cost of biodiesel would be twice that of current petroleum diesel fuel. In particular, the theoretical estimation of 30-50% conversion of microalgal biomass to oil, giving an oil yield of 47,000 to 308,000 L.h^{-1} y^{-1} (Briggs 2004; Vincent Inc.2007) or 160t.h^{-1}y^{-1}(Rapier, 2008) seems overly generous. None of the projects so far reported have come anywhere close to achieving these oil yields (Dimitrov 2007) and the estimated costs are always well in excess of current market prices per barrel. Other factors to be considered relate to the high demand of microalgal-based biomass systems for water and nutrients, the problems of ensuring optimal light supply and the mechanics and cost of harvesting. The best microalgal biomass yield under tropical conditions is reported to be ca. 50t.h^{-1}.yr^{-1}(Biopact 2007), less than 10% of the theoretical maximum (Vincent Inc.2007). Production of biofuels from microalgal biomass, despite its obvious appeal, presents many challenges that would have to be overcome before it could be employed on a large scale. It would be most economically feasible when combined with other processes such as the production of bulk chemicals, food and feed ingredients or bioremediation and waste treatment systems.

Chapter IX

Hydrogen-Producing Microalgae

In view of the known importance and increasing use of molecular hydrogen as a fuel, it is not surprising that the search for alternative, environmentally acceptable and renewable sources of this valuable source of energy is now receiving considerable attention. Inevitably, the fact that certain microalgae have long been known to carry out reactions having molecular hydrogen as a product (e.g Gaffron & Rubin 1942) raises interesting possibilities. Any consideration of the hydrogen metabolism of microalgae focuses primarily on two key groups of enzymes namely the hydrogenases that catalyse the reversible reaction $H_2 \leftrightarrow 2H^+ + 2e^-$ and the nitrogenases that, in nitrogen-fixing cyanobacteria, catalyse the reduction of nitrogen to ammonia but also yield hydrogen as a by-product. The hydrogenase and nitrogenase enzymes have in common their extreme sensitivity to oxygen, a feature that has to be overcome, given that in photosynthetic microalgae (and cyanobacteria), unlike the photosynthetic bacteria (which also have hydrogenase activity) (Llama et al. 1981), the most common source of reducing power is the photochemical splitting of the water molecule. Among photosynthetic eukaryotes, the ability to produce hydrogen is a unique feature of certain microalgal species.

The Production of Hydrogen by Cyanobacterial Nitrogenase Systems

This process occurs in those relatively small number of cyanobacterial species that are capable of fixing nitrogen by reducing elemental nitrogen to the oxido-reduction level of ammonia (NH_3). In these nitrogen-fixing cyanobacteria the nitrogenase enzymes that catalyse reduction of N_2 to NH_3 also produce hydrogen in what has been described as a seemingly wasteful process (Prince & Kheshgi 2005), since the evolved hydrogen is no longer available for nitrogen fixation. The amount of hydrogen "lost" varies between different types of nitrogenase enzyme. Those having molybdenum as a component of the active site are the most "efficient" and the reactions involved follow the same pattern as those described for the nitrogen-fixing bacterium *Azotobacter vinelandii* (Eady, 2003) which can be summarised as follows:

$$N_2 + 8H^+ + 8e^- = 2NH_3 + 1H_2$$

According to this reaction, for every molecule of nitrogen fixed, one molecule of hydrogen is released. But this exact relationship between hydrogen evolution and nitrogen fixation holds only under certain appropriate conditions. Under conditions of low nitrogen availability, hydrogen is released at a higher rate while under high nitrogen availability, less hydrogen is released. Thus, under conditions of restricted delivery of nitrogen, some cyanobacteria can increase the formation of hydrogen by 2.5 to 3.5 times compared with that produced under normal nitrogen supply, with maximum hydrogen yields up to 40ml.g dry cells^{-1}.h^{-1} (reported in Varfolomeev & Wasserman 2011). Under conditions favourable for growth, however, cyanobacteria normally release negligible amounts of hydrogen through this mechanism.

Cyanobacterial nitrogenases may contain molybdenum (Mo), vanadium (V) or iron (Fe) in the active site (Eady 1996). Under conditions of molybdenum deprivation, the molybdenym nitrogenase is replaced by an alternative vanadium nitrogenase and if V is also limited, some nitrogen-fixing micro-organisms are able to synthesise an alternative iron-nitrogenase (Tamagnini et al. 2007).

Forms of nitrogenase having vanadium in the active site release three molecules of hydrogen for every molecule of nitrogen fixed as follows (Eady 1996):

$$N_2 + 12\ H^+ + 12e^- = 2NH_3 + 3H_2$$

Cyanobacterial V-nitrogenase was found by Masukawa et al. (1998) to be more sensitive to oxygen than the Mo-nitrogenase. Iron nitrogenase enzymes are the least efficient nitrogen fixers of the nitrogenases, releasing 7.5 molecules of hydrogen for every molecule of nitrogen fixed, as follows:

$$N_2 + 21H^+ + 21e^- = 2NH_3 + 7.5H_2$$

Production of hydrogen by any of the above reactions would require a substantial input of metabolic energy. The energy requirement of the reaction catalysed by Mo-nitrogenase, for example, can be summarised as follows:-

$$N_2 + 8H^+ + 8e^- + 16ATP = 2NH_3 + 16ADP + 16\ P_i + H_2$$

with a requirement for hydrolysis of approximately 2 molecules of ATP for every electron contributing to hydrogen production (Eady 1996).

In addition to the nitrogen enzyme, cyanobacteria also possess hydrogenase enzymes that catalyse the reaction:

$$2H^+ + 2e^- = H_2$$

which, when proceeding in the "uptake" direction, consumes H_2 so reducing the net yield of hydrogen (but possibly making nitrogen fixation more energy efficient). Cyanobacterial NiFe hydrogenase enzymes may be strictly of the uptake type or bidirectional and thus able to catalyse the uptake or release of hydrogen, depending on the conditions (Tamagnini et al. 2007). The link between H_2 production and N_2 fixation in cyanobacteria became evident when it was observed that both H_2 production and N_2 –fixation (as measured by $C_2\ H_2$ reduction) decreased in parallel if fixed nitrogen was added to the culture medium (Adams et al. 1981). Additionally, it was shown that H_2 production is light-dependent (PSI only), is sensitive to uncouplers of photophosphorylation and is inhibitated by N_2

The nitrogenase enzyme is very sensitive to molecular oxygen and is only active under anaerobic conditions. Thus, cultures of *Anabaena cylindrica* release significant amounts of hydrogen for a prolonged period only when maintained in strictly anaerobic conditions, while growing cultures of *Anabaena variabilis* and *A. azollae* did not produce hydrogen when aerated with air+CO_2 or argon + CO_2 but did so when aerated with argon alone (Tsyganov et al. 1998).

In heterocystous cyanobacteria grown in air without combined nitrogen, the uptake hydrogenase, a membrane-bound enzyme (Houchins 1984) is mainly confined to the heterocysts where it is protected from oxygen inactivation. Its physiological function is to regain H_2 /electrons produced during the nitrogenase reaction (Tamagnini et al. 2007). It removes oxygen from the vicinity of the nitrogenase thereby protecting it from inactivation while at the same time supplying reducing equivalents (electrons) to various cell functions. It also minimizes loss of energy by contributing ATP via the oxy-hydrogen reaction.

The bidirectional hydrogenases of cyanobacteria have been described as being analogous to bacterial hydrogenases (Volbeda et al. 1995). Under anaerobic conditions, they produce and evolve significant amounts of H_2 (see below) in a reaction that does not require ATP. It is therefore more energy-efficient than nitrogenase-based H_2 production but, since the enzyme is not located in an oxygen-protected environment, the reaction tends to go in the opposite direction (i.e. H_2 uptake) when H_2 partial pressure is above a certain critical level (Tamagnini et al. 2007). For the bidirectional hydrogenase to function in H_2 production would require continuous removal of H_2 (and O_2).

The link between photosynthesis and hydrogen production and nitrogen fixation by cyanobacteria is summarized in Figure 12, where $PSII^*$ and PSI^* represent the excited states of the two photosystems and the arrows show the flow of electrons through various intermediate carriers to $NADP^+$. Under the appropriate conditions, hydrogenase or nitrogenase enzymes would divert electrons from the fixation of CO_2 to hydrogen production or nitrate reduction, respectively. Of these two enzymes, nitrogenase has been described by Prince & Kheshgi (2005) as having serious deficiencies as a potentially efficient producer of hydrogen. In addition to its high energy (ATP) requirement, nitrogenase has a relatively weak catalytic activity and, because of its extreme sensitivity to oxygen, requires special protective mechanisms such as is afforded in some nitrogen-fixing cyanobacteria by its localisation in heterocysts which lack PSII (Golden & Yoon 2003).

Nitrogenase of cyanobacteria is most active under conditions of deprivation of combined forms of nitrogen (nitrate, nitrite, ammonia or urea). It is likely, therefore, that the conditions that allow nitrogenase activity (and hence hydrogen production) would be far less favourable for good yield and quality of proteins (e.g. as livestock feed). To test for this, Wünschiers (2006) compared the nitrogen content of proteins produced by the filamentous cyanobacterium *Nostoc* (PCC 7120) under nitrogen-fixing and non-nitrogen-fixing conditions. It was shown that when growing on dinitrogen, *Nostoc* actually incorporates more nitrogen into its proteome than it does when grown on ammonia. This seemingly paradoxical observation would clearly have interesting biotechnological implications in relation to the phohotobiological production of hydrogen via the nitrogenase reaction, although the high metabolic cost would remain as a considerable barrier.

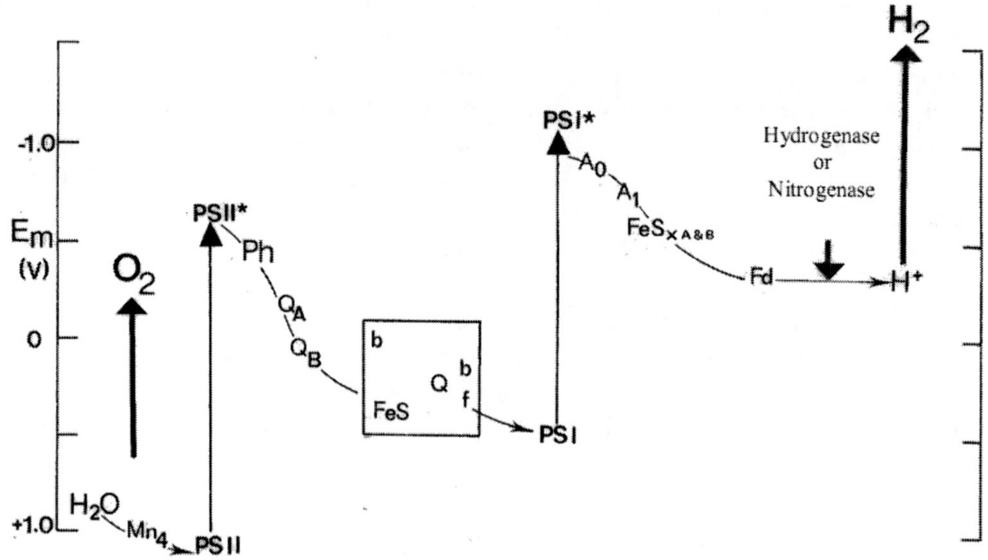

Figure 12. Energetics of electron transfer from the excited states of the two respective photosystems PSII* and PSI* showing some of the key intermediate electron carriers during the direct photoproduction of hydrogen and oxygen through the action of hydrogenase or nitrogenase. (Ph – phaeophytin, Q – quinones A and B, The box – cytochrome b_6f complex, FeS_x iron-sulfur clusters, Fd – ferredoxin). (Redrawn, with permission, from Prince & Kheshgi, 2005).

Some of the specific challenges that would need to be met before cyanobacterial hydrogenases could achieve a sustained yield of hydrogen have been summarised by Tamagnini et al. (2007). It would be necessary, for example, to maintain conditions that direct energy flow through nitrogenase towards H_2 production rather N_2 fixation and, if possible, to enhance over-expression of nitrogenase and/or bidirectional hydrogenase (e.g. through an increase of heterocyst frequency). Another possible strategy would be to induce expression of the most active H_2 producer (Fe-hydrogenase) in heterocysts. The oxygen sensitivity of nitrogenase and/or hydrogenase might perhaps be overcome by transferring an O_2 –tolerant bacterial NiFe hydrogenase into a cyanobacterial cell. The feed-back effect of accumulating ATP (not used in N_2 fixation) would need to be addressed as also would the low quantum efficiency of a system based on PSII and PSI antennae.

Production of Hydrogen during Cyanobacterial Fermentation

Cyanobacteria are essentially aerobic organisms having a respiratory metabolism during which endogenous storage products of photosynthesis are oxidised to produce energy. In the absence of oxygen, a number of cyanobacteria have been shown to undergo a fermentation process during which a range of different substrates may be utilized (Stal & Moezelaar 1997). Conditions that allow cyanobacterial fermentation are also favourable for reactions leading to the production of hydrogen, catalyzed by a hydrogenase enzyme resembling that known for many chemoorganotrophic bacteria where its function is presumed to serve to release excess reductant (Thauer et al. 1977). Cyanobacterial fermentation most commonly occurs in the dark and hence in the absence of photosynthetically produced oxygen. It occurs, for example, in the anoxic hypolimnion of a strongly stratified lake such as. Solar Lake, (a hypersaline pond in the Sinai desert) where blooms of *Oscillatoria* sp. have been shown to carry out an anoxic form of photosynthesis during which sulphide acts as electron donor, becoming oxidised to elemental sulphur (Cohen et al. 1977) and energy is generated by anaerobic fermentation of glycogen (Oren & Shilo 1979). Cyanobacterial fermentation can also occur, at least periodically, in microbial mats and in laminated microbial sediments where restricted diffusion of oxygen results in a build-up of anoxia (Revsbech et al. 1983).

In all these cases, the switch from photoautotrophy to fermentation is immediate, suggesting that the fermentation enzymes are constitutive and do not involve induction of new enzyme proteins. Indeed, fermentation can also occur under reduced partial pressure of oxygen (Heyer et al. 1989) or even under atmospheric oxygen partial pressure when aerobic respiration is inhibited by cyanide (Oren & Shilo 1979). The activity of the enzymes of the fermentation pathway is, however, much lower than that of the enzymes of aerobic respiration or of phototrophic metabolism (Stal & Moezelaar 1997).

In cyanobacteria therefore, hydrogenase can be regarded as a constitutive enzyme that is active only under the appropriate conditions of anoxia and in the presence of an alternative source of energy from fermentation. In *Anabaena cylindrica,* for example, hydrogenase activity can be detected after 1-5 h of dark anaerobic incubation (Hallenbeck et al. 1981). The enzyme catalyses the reversible activation of hydrogen and was shown by Houchins (1984) to play an important role in cyanobacterial metabolism generally. It is an NiFe enzyme of the bidirectional type that probably plays a role in fermentation and/or acts as an electron valve during photosynthesis.

Hydrogenases of Eukaryotic Microalgae

The earlier reports of metabolic hydrogen production by microalgae, (e.g. Spruit 1958), envisaged a hydrogenase enzyme that is inactive under aerobic conditions but which becomes activated after a period of anaerobiosis. The enzyme, although temporarily inactive in the presence of photosynthetically produced oxygen, rapidly recovers its activity as soon as the oxygen is removed. The observation by Kessler (1956) that anaerobically adapted cells of *Ankistrodesmus braunii* were able to reduce nitrite to the level of ammonia:

$$HNO_2 + 3H_2 \rightarrow NH_3 + 2H_2$$

pointed to a possible role for hydrogenase in microalgae generally and its possible involvement in the disposal of excess electrons. The early work of Gaffron & Rubin (1942) described hydrogen evolution by adapted microalgal cells as occurring in two separate processes, a fermentative process in the dark and a much more rapid photochemical process in the light. Electrochemical methods of detection confirmed that both hydrogen and oxygen were at first formed simultaneously by *Chlorella* cells illuminated under anaerobic conditions, with water as the probable common source (Spruit 1958). Hydrogen production ceased, however, as soon as the concentration of oxygen exceeds a critical (quite low) value.

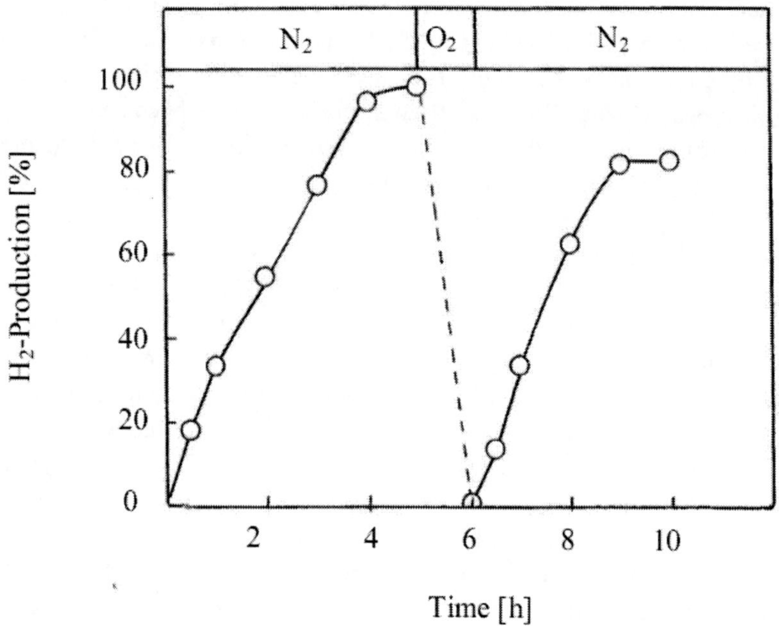

Figure 13. Activation, inactivation and reactivation of *in vivo* hydrogenase activity of *Scenedesmus obliquus* under alternating anaerobic (N_2) and aerobic (O_2) conditions. (Redrawn, with permission, from Wünschiers et al. 1998).

The sensitivity of microalgal hydrogenase to oxygen was confirmed by Wünschiers et al. (1998) for the green alga *Scenedesmus obliquus* (see Figure 13). After adaptation of the cells to anaerobic conditions, full hydrogenase activity was shown to develop within about four hours. Exposure of the cells to mere traces of oxygen resulted in complete loss of hydrogenase activity within minutes but was rapidly restored when anaerobic conditions were reimposed. The sensitivity to oxygen was shown to be due to an effect of oxygen on the FeS centres of the enzyme (Adams 1990).

Hydrogen production by eukaryotic microalgae was shown by Stuart & Kaltwasser (1970) to be insensitive to DCMU leading them to suggest that Photosystem II was probably not involved, a conclusion confirmed by work with mutants lacking PSII (Stuart & Gaffron 1972). To prevent the oxygen-induced inactivation of hydrogenase in suspensions of the green microalgal *Scenedesmus obliquus*, Randt & Senger (1985) used sodium dithionite as an oxygen scavenger. They were able thus to show that in a number of different mutants, photoproduction of H_2 could occur by different mechanisms depending on the level of PSII activity (or its complete absence). What they described as "core" H_2 – photoproduction occurred in mutants having only PSI activity. Under conditions allowing some development of PSII activity (e.g. re-greening of heterotrophically grown cells) there was an additional H_2 –photoproduction activity. The core H_2 production, it was concluded, is dependent upon PSI only and can use sources other than water as electron donors. The "non-core" H_2 – photoproduction process is dependent upon a functioning PSII and requires water as electron donor. Cells of *S. obliquus* in which PSII is blocked gain their reducing power through fermentation of organic substrates.

The energetics of H_2 – photoproduction by eukaryotic microalgae (and cyanobacteria), compared with that which might apply in the case of green or purple bacteria, is summarised in Figure 12. Based on such considerations, Wünschiers et al. (1998) noted that the reducing power of the first stable intermediate (ferredoxin) of photosystem I in cyanobacteria and the photosystem I homolog in green bacteria should be sufficient to allow reduction of protons by hydrogenases. The reducing power of the first stable intermediate (quinone) of photosystem II, however, and its homolog in purple bacteria, is not sufficient to produce hydrogen.

The maximum quantum efficiency for photoproduction of H_2 in a number of green microalgae has been reported as 0.21 – 0.24 mol H_2. mol^{-1} absorbed quanta. The quantum yield of H_2 (20-25%), as reported by Boĭchenko & Hoffmann (1994), is close to the theoretical value for the two-electron reaction of H_2 production from a PSI:PSII ratio of 1:1. Hydrogen production by eukaryotic microalgae can occur without ATP input and the hydrogenase enzyme is exceptionally active with turnover numbers approaching 10^6 s^{-1} (Prince & Kheshgi 2005). This contrasts with cyanobacterial nitrogenases which require at least 4 mol. of ATP for every mol. of H_2 produced. This makes the hydrogenase-mediated reaction linked to PSI four times more efficient relative to the number of photochemical centres than the ATP-dependent H_2 evolution *via* nitrogenase in photosynthetic bacteria and cyanobacteria (Warthmann et al. 1993).

The energetic efficiencies of various forms of photobiological production of hydrogen were assessed by Prince & Kheshgi (2005). They calculated the energy requirement to drive 4 electrons through the photosynthetic electron transfer system in the hydrogenase reaction:

$$2H_2O = 2H_2 + O_2 \ \Delta H = 572.5 \text{ kJ}$$

and concluded that for light at 680 nm the 8 photons would yield 1405 kJ, giving a maximum energy efficiency of 40.7%. Hydrogen production by the nitrogenase reaction, since it requires an additional minimum of 2ATPs to drive each electron through the system, would substantially reduce the efficiency of the process – to an estimated 20.3% quantum efficiency for light at 680 nm. If hydrogen production occurs via an indirect route involving hydrolysis of previously synthesised photosynthetic product, the maximum quantum efficiency using 680 nm light is 27.1% whereas for the equivalent process using nitrogenas it drops to 16.3%.

Characterisation of the Hydrogenase of Eukaryotic Microalgae

The first eukaryotic microalga to yield cell-free preparations of hydrogenase was the chlorophyte *Chlamydomonas reinhardtii* (Abeles 1964) and the enzyme from the same species was later partially purified by Roessler & Lien (1984). Then Happe & Naber (1993) isolated and purified the hydrogenase from cells anaerobically adapted by flushing with argon for 2-4.5h in the dark. All procedures were performed under strictly anaerobic conditions and protein precipitations carried out in the presence of 200mM sodium dithionite. Purification by successive column chromatography steps also required anaerobic conditions.

The purified enzyme had an apparent molecular mass of 48 kDa with 4 Fe atoms/mol and a specific activity for H_2 evolution of 935 μmol.min^{-1} mg $protein^{-1}$ at 25°C, using methyl viologen as electron donor. Based on an examination of the amino acid sequence of an identified fragment of the enzyme protein, it was concluded that the *C. reinhardtii* hydrogenase is not homologous with a number of other prokaryotic Fe-hydrogenases. These observed structural dissimilarities led Happe & Naber (1993) to conclude that the *Chlamydomonas* enzyme may belong to what was then considered to be a new class of hydrogenases different from those of bacteria. The differences, they further suggested, may reflect differences in function.

The hydrogenase of another chlorophyte, *Scenedesmus obliquus,* was described as being a nickel-iron enzyme whose protein component consisted of two subunits of ca.36 and 55 kDa (Schnackenberg et al. 1993). It was further characterised as a new type of iron hydrogenase (HydA) with a molecular mass of 44.5 kDa (Florin et al. 2001). It reacted with antibodies raised against the iron hydrogenase from *Chlamydomonas reinhardtii* and its transcription was induced during anaerobic incubation. The *S. obliquus* HydA enzyme was confirmed as belonging to the class of iron hydrogenases known from other organisms but with some structural features probably related to a novel pathway of electron transfer between the enzyme and the photosynthetic electron transport chain.

Of the three types of hydrogenases, the [FeFe] type, the [NiFe] type and the FeS-cluster-free (Fe-only) type, the first has the highest turover rate (Frey 2003) and is the most active in microalgal hydrogen production. The different forms of Fe-hydrogenases are considered to be phylogenetically distinct but with strong similarities between their protein components. The active site invariably contains CO and CN ligands and a binuclear metal centre. In general, [NiFe] hydrogenases catalyze H_2 uptake whereas [Fe] hydrogenases tend to catalyze H_2 evolution (Frey 2001). A likely structure for the latter enzyme and its catalytic site has been proposed by Ghirardi (2006) who has also discussed the pathway by which H_2 gas leaves the catalytic site. The catalytic site of the [Fe] hydrogenases consists of a [4Fe4S] cluster connected through a bridging cysteinyl ligand to a binuclear [2Fe] centre (Peters et al. 1998).

The genes encoding the two protein components (HydA1 and HydA2) of the enzyme are described as being nuclear and the proteins themselves are deemed to contain putative transit sequences that target them to the chloroplast. Transcription of the hydrogenase genes requires anaerobic conditions (Forestier et al. 2003) and inactivation of [FeFe]-hydrogenase is believed to occur by irreversible binding of O_2 to the H-cluster embedded within the enzyme structure. The genes necessary for photoproduction of H_2 in *Chlamydomonas reinhardtii* were identified by Posewitz et al. (2004) by screening random insertional mutants for colonies unable to produce the gas. They were identified as *HydEF* and *HydG* and shown to be transcribed anaerobically in parallel with previously described *HydA1* and *HydA2* [Fe] hydrogenase genes of *C. reinhardtii.* Disruption of *HydEF*, they showed, abolished all H_2 production although full-length hydrogenase protein was still detected. Hydrogen production was restored after complementation of the *HydEF-1* mutant with DNA containing the *HydEF* gene.

Several *C. reinhardtii* mutants, compromised in their ability to photoproduce H_2, were identified by Posewitz et al. (2004), one of which (*sta7-10*), revealed a disruption of a gene with high homology to the isoamylase family of enzymes known to be important for the accumulation of crystalline starch in a variety of photosynthetic organisms. This apparent link between hydrogen production and starch metabolism was confirmed when complementation of mutant cells with genomic DNA corresponding to the *STA7* gene was found to restore both starch accumulation and H_2 production. These, and other related observations, were interpreted to indicate that anaerobiosis, on its own, is not sufficient to maintain hydrogenase gene expression. The underlying physiology, of which starch metabolism is an important component, must also be intact. The authors speculate that starch degradation affects the supply of reducing power (NADPH) and/or the oxidation state of the plastoquinone pool, both of which are known to regulate gene transduction.

The active form of hydrogenase of eukaryotic microalgae has been shown to be exclusively located in the chloroplast (e.g. Maione & Gibbs 1986) in amounts comparable with the number of photosynthetic reaction centres (Boĭchenko & Hoffman 1994). It has further been suggested that the enzyme is probably loosely bound to the thylakoid membrane, with the two reducing equivalents necessary to produce a molecule of H_2 being generated by a single turnover of the photosynthetic reaction centres. One molecule of hydrogenase is visualised as co-operating with the two centres *via* a common pool of soluble ferredoxin. Since most of the algae investigated are also able to effect dark ATP-dependent H_2 evolution (Klein & Bet 1978), it is likely that hydrogenase may also be sustained in a stable semi-reduced state through links to fermentative metabolism.

The immediate source of electrons for PSI-driven H_2 photoproduction is the pool of reduced carriers between the two photosystems, as is shown in Figure 12. Depending on irradiance, the PSI donor pool is competitively re-reduced by an electron flow from fermentative metabolism, or other sources including the PSII-driven electron flux from the water oxidising system (Boĭchenko & Hoffmann 1994). During the first seconds after onset of saturating irradiation, the PSII-driven electron input is negligible, thus the turnover of PSI-mediated hydrogen production is limited mainly by the electron input from fermentative metabolism.

Practical Considerations of Bulk Photobiological Hydrogen Production by Microalgae

One positive feature of the production of biohydrogen by microalgae is that the hydrogen does not accumulate in the biomass but is quickly released in the gas phase. Set against this, as far as photosynthetic microalgae are converned, is the extreme sensitivity of the processes to oxygen which reacts with the FeS centres of the two enzymes involved. In heterocystous nitrogen fixing cyanobacteria, nitrogen fixation (and the potential location of H_2 production) is physically isolated from the oxygen-producing reactions of photosynthesis, the former in the heterocysts; the latter in the vegetative cells. In his survey of the problems that have to be overcome in any attempt to achieve H_2 production by microalgal biophotolysis, Benemann (2000) refers to the very high metabolic cost of the heterocyst-based reaction. In addition to the 4ATP per H_2 evolved, there is the further cost of synthesis and maintenance of the heterocysts (Turpin et al. 1985) and the cost of separating the two products, H_2 and O_2. The increased requirement for ATP might be satisfied by greater starch mobilisation, perhaps through the selection of new mutant strains with increased reserve starch content (Kruse et al. 2005).

Melis et al. (2000) achieved a temporal separation of H_2 and O_2 production, with nitrogen fixation (and hence H_2 production) confined to periods of anaerobic dark fermentation alternating with periods of light-dependent photosynthesis. Yet another option that has been considered (Miyamoto et al. 1987) would be the use of mixed cultures of microalgae and photosynthetic bacteria. In such a system the microalgal component would fix CO_2 and the photosynthetic bacteria would produce H_2. A small pilot plant based on this concept used two $2m^2$ open ponds to grow cultures of *Chlamydomonas reinhardtii* to provide substrate for feeding to closed bacterial photobioreactors via an intermediate dark fermentation stage. Such a set-up, using a hollow fibre ultrafiltration unit to separate the microalgal fermentation products for feeding to the photosynthetic bacterial photobioreactors, achieved continuous operation, yielding several litres of H_2 per day (Ikuta et al. 1998). The relative inefficiency of bacterial photosynthesis was, however, deemed to impose serious limitations.

The realdrawback with any system using nitrogenase as the key enzyme for H_2 production is its inefficiency and its high energy (ATP) requirement. Systems based on reversible hydrogenase would appear to be much more promising. The specific H_2 -evolving capacity of hydrogenase is almost a thousand times greater than that of nitrogenase (Benemann 2000) and, most importantly, does not require ATP. Reversible hydrogenase is, however generally expressed at relatively low activities (Benemann 2000) with consequent much lower actual H_2 production rates. Moreover, H_2 production by reversible hydrogenase is typically inhibited by $<10\%$ partial press of H_2. There is also a requirement for a store of carohydrate substrate to support the cells under anaerobic conditions to stimulate hydrogenase activity.

Attempts have been made to overcome the sensitivity to oxygen of the hydrogenase enzyme by the use of O_2-absorbing agents to keep the partial pressure of oxygen low. In a pilot reactor using anaerobically adapted cultures of *Scenedesmus obliquus* and supplying Na dithionite to recapture the photosynthetically produced oxygen, Wunschiers et al. (1998) achieved a H_2 production rate of at least 1ml/hr by 30g algae (wet weight). Gas chromatographic analysis of the gas produced showed it to be pure (uncontaminated) hydrogen which, when injected into a polymer electrolyte membrane fuel cell, produced enough electric power to drive a small electric motor.

Wykoff et al. (1998) explored the possibility of using sulphate deprivation to inhibit the oxygen evolving component of photosynthesis whilst allowing continued H_2 production. Cultures of *Chlamydomonas reinhardtii* deprived of sulphate exhibited substantial inhibition of oxygen evolution. One day of S starvation was reported to cause a 75% decline of O_2 evolution without significantly affecting the light-saturated rate of photosystem I (PSI) activity. Light-saturated PSII activity, on the other hand, was reduced by more than 50% due, it was suggested, to loss of active PSII centres. Under these conditions, oxidative respiration by the cells depletes any residual O_2 and the cultures become anaerobic (Melis et al. 2000). This resulted in induction of hydrogenase and sustained H_2 production in the light, but not in the dark. It was clear that light absorption by PSI was required for H_2 evolution indicating that photoreduction of ferredoxin is followed by electron donation to the hydrogenase. The induced enzyme then catalyses the reduction of protons to molecular hydrogen in the chloroplast stroma (Melis et al. 2000). The hydrogen is produced through photo-oxidation of water (Zhang & Melis 2002) and is presumed to be the only mechanism available for generating the energy (ATP) required for survival under these S-depleted anaerobic conditions. In sulphur-deprived batch cultures of *C. reinhardtii,* photoproduction of H_2 was reported to last for 100-150h but could be reinitiated by allowing recovery of the culture for a two-day period in a sulphate-sufficient medium before imposing a second period of sulphur deprivation (Ghirardi et al. 2000a).

Under conditions of sulphur limitation, there are two main pathways of electron supply, a direct and an indirect pathway. In the former, hydrogen is produced using electrons supplied from the photosynthetic electron transport chain (pet) from PSII to PSI and then to the hydrogenase *via* reduced ferredoxin. In the latter pathway, the reducing power generated by oxygenic photosynthesis (which, under sulphur starvation is stored as starch) is regenerated during starch catabolism and injected into the intersystem pet chain at the level of the plastoquinone (PQ) pool. The enzyme involved in this process is a plastidial type II NADH dehydrogenase Nda2 (Jans et al. 2008).

Sulphate deprivation interferes with the high photosynthetic electron flow to ferredoxin. It also leads to an initial over-accumulation of starch and to the development of anaerobiosis because sulphate deprivation has only a slight effect on respiration (Melis et al. 2000). The rate of H_2 production under sulphate deprivation is only about $1/10^{th}$ of the capacity of the photosynthetic apparatus to process electrons (Ghirardi et al. 2005) and only lasts for 3-4 days of sulphur deprivation. Ghirardi et al. (2000a) proposed running cycles of 5 days in minus sulphate alternating with periods of 2 days in plus sulphate. Kosourov et al. (2003) investigated how the yield of H_2 from sulphur-deprived cultures of *C. reinhardtii* was affected by extracellular pH and noted that maximum yield was obtained at pH 7.7. They further observed that the pH profile of H_2 photoproduction corresponded with that of the residual PSII activity, but not with the pH profile of photosynthetic electron transfer through PSI or of starch and protein degradation. By cantrolling the pH, they concluded, it should be possible to facilitate high H_2 photoproduction and low yields of fermentation.

Federov et al. (2005) hypothesised that the H_2 – producing phase could perhaps be prolonged by continuously replacing the cells during the H_2 – producing phase with sulphur-deprived but photosynthetically active cells. To test this, they devised a two-stage chemostat system that physically separated photosynthetic growth of *C. reinhardtii* cultures from H_2 production. It comprised two automated photobioreactor chambers in one of which the cultures were maintained in chemostat mode under sulphate limitation to yield photosynthetically competent cells. These were continuously delivered to a second photobioreactor maintained under anaerobic conditions where H_2 production could occur. It was shown that the cells grown in the sulphate-limited chemostst achieved steady-state growth rates controlled by the rate of supply of sulphate, with negligible extracellular residual sulphate in the medium. Upon transfer to the second bioreactor chamber, the cells quickly became sulphur-deprived and hence capable, under the prevailing anaerobic conditions, of long-term (>4,000h) H_2 production. Maximal H_2 photoproduction rates of 0.58 and 0.42 mlh^{-1} were reported in cultures with 60μM sulphate input at two dilution rates and with 50 $\mu E.m^{-2}.s^{-1}$ illumination.

Laurinavichene et al. (2008) devised a culture system in which the microalgal cells of a non-motile mutant of *Chlamydomonas reinhardtii*, were immobilized on a glass fibre matrix under argon flow within the photobioreactor and fed with a continuously flowing medium with regulated supply of sulphate. Decreasing the sulphate concentration in the flow medium (to 20μM) 2days after the start of the experiment resulted in an initial increase in the rate of H_2 photoproduction, to reach a peak of $11mld^{-1}$ on days 8-9. Thereafter, the rate of H_2 production decreased to reach a steady level of 4-6ml day^{-1} which lasted for at least 90 days despite a further decrease of sulphate supply (to 10μM) on day 16. In another experiment it was shown that periodic additions of sulphate to the flowing medium could restore H_2 production with rates oscillating between lows of <1ml.day^{-1} to highs up to >10ml.day^{-1}. Although immobilisation of microalgal cells on to glass fibres tended to reduce the specific rate of H_2 production by 20-25%, the period of H_2 production could be increased from 3-4 days to 21 days.

These (and other similar findings) suggested the possibility of effecting a temporal separation of the O_2 evolution and H_2 producing processes, using substrate availability as a reversible switch to regulate activity of the O_2 –evolving PSII complex. The reversibility of the method was confirmed by cycling a *C. reinhardtii* culture between the two stages (oxygenic photosynthesis and H_2 production) for up to three full cycles. Melis et al. (2000) also offered an explanation for the observed evolution of H_2, suggesting that it is the only mechanism available, under the prevailing S-depleted anaerobic conditions and hence in the absence of mitochondrial respiration and oxygenic photosynthesis, capable of generating the energy (ATP) necessary for survival.

The process of H_2 production, it was observed, entails protein consumption and electron transport from endogenous substrate to the cytochrome b_6-f and PSI complex in the chloroplast thylakoids. The establishment of anaerobiosis by S deprivation is an energy-dependent process that requires a carbon substrate for respiration. Melis et al. (2000) showed that the main substrate for respiration over the initial 30h of S-deprivation was acetate but once anaerobiosis has been established acetate is no longer utilized – its main role in establishing anaerobiosis having been accomplished. These studies were the first to show that it is possible to produce and accumulate significant volumes of H_2 gas in a sustainable photobiological process that can be maintained for several days.

Based on current understanding of the biology of hydrogen production by microalgae, the most feasible approach to possible commercial application would be to combine large-scale, low-cost production of microalgal biomass with more sophisticated bioreactor technology for hydrogen production and capture.

One of the major problems to be overcome, assuming a location providing adequate year-round solar irradiance and assuming a most optimistic 10% efficiency of conversion of solar energy to hydrogen, would be the vast areas of land that would be required to display the photobioreactor component to best effect. Price and Kheshgi (2005) have estimated that to provide the hydrogen equivalent of 806,000 gallons (US) of gasoline per year (the average annual output of only one of the country's many thousands of. gasoline station) would require a photobioreactor the equivalent of 147,000 m^2. Allowing for pre-photobioreactor mass culture of microalgal biomass in open ponds would clearly further increase the area of land required. Another technical problem that would have to be solved relates to the mass transfer of hydrogen gas out of the photobioreactor's liquid phase and into storage. Greenbaum (1988) has suggested flushing the hydrogen (and oxygen) from the system using an inert gas, a process demanding careful handling given the extremely dangerous nature of this mix of gases.

Geier et al. (2012) investigated the sustainability of a two-stage process in which both cultivation and subsequent hydrogen production by *Chlamydomonas reinhardtii* occurred in natural (sunlight) conditions.

The cells were first grown in closed photobioreactors under simulated outdoor conditions (up to 2,000 μmol photons.m^{-2} s^{-1}) with additional CO_2 supply (10%) before transfer to the anaerobic conditions for hydrogen production in a sulfate-free medium. However, the yield of hydrogen during the second stage fell well short of that achieved from cells grown under controlled laboratory conditions.

Possible Future Developments

One of the major impediments to the commercial production of hydrogen from microalgal hydrogenase systems is the sensitivity of the enzyme to oxygen. Attempts to overcome this problem include the employment of mutagenic procedures in a search for mutants capable of maintaining hydrogenase activity in the presence of specified amounts of oxygen Ghirardi et al. (1998). Using ethylmethane sulfonate as a mutagen, a large number of mutant clones of *Chlamydomonas reinhardtii* were produced and then screened for their ability to evolve hydrogen in the presence (or absence) of oxygen. All the clones tested exhibited vigorous H_2 evolution in the absence of oxygen but none displayed any significant tolerance of their hydrogenase to oxygen. Another possible approach suggested by Ghirardi (2006) was to identifying the pathways by which O_2 gains access to the catalytic site of the hydrogenase enzyme, and then to attempt to reduce such accessibility (by mutagenesis). The diffusion pathway for O_2 in the [FeFe]-hydrogenase of *Clostridium pasteurianum* is known to occur along two pathways (Cohen et al. 2005). The next step would be to identify these pathways in microalgae with a view to manipulating them to restrict the flow of oxygen without concomitantly affecting H_2 gas diffusion properties.

Attempts have also been made to modify the hydrogenase enzyme in such a way as to make it less sensitive to inhibition by oxygen. Site-directed mutagenetic techniques have succeeded in rendering the hydrogenase reaction in the bacterium *Azotobacter vinelandii* insensitive to oxygen inhibition albeit with substantial loss of hydrogen production. It might therefore be possible to replace the microlagal hydrogenase with the equivalent, less oxygen-sensitive, bacterial enzyme. The normal fate of photosynthetically generated reductants is the ferredoxin/NADPH oxidoreductase (FNR) that generates NADPH for other metabolic pathways such as the enzymes of the CO_2 fixation cycle. Any O_2 tolerant hydrogenase would, therefore, have to compete for reductants with FNR which is known to have a higher affinity for ferredoxin than does hydrogenase.

Recognising that [2Fe-2S]ferredoxin represents a critical branch point in the flow of electrons from Photosystem I towards different metabolic pathways (including proton reduction by hydrogenases), attention has naturally been directed towards the primary competitor with FeFe hydrogenase for low potential electrons under physiological, aerobic conditions, namely ferredoxin-NADP$^+$oxidoreductase (FNR) (Knaff 1996). Studies involving chemical modifications have shown a close interaction between ferredoxin and FNR in *Chlamydomonas reinhardtii* suggesting that possible future metabolic engineering efforts should be directed towards the molecular interactions between the FeFe hydrogenase isoenzyme Hyd A_2 and [2Fe-S] ferredoxin (Chang et al. 2007). Such studies, it was suggested, might lead to potential mutagenic strategies to modulate electron flow to HydA_2

Noting that the hydrogenase of photosynthetic bacteria contains only a single 4Fe-cluster, Adams et al. (1981) argued that the more complex, high molecular weight structures of microalgal hydrogenase systems may not be an obligatory requirement for operation of the basic reversible H_2 activation process. On this basis, it was speculated that it may be possible to develop a functional synthetic model hydrogenase systems composed of Fe_4S_4 clusters with oligopeptide ligands to mediate electron transfer to hydrogenase. The technical problems involved in the development of such a purely synthetic hydrogenase would clearly be considerable. Better, it was suggested, to focus on optimising the conditions for *in situ* functioning of the microlagal hydrogenase.

Chen & Melis (2004) have identified a sulphate permease gene which is responsible for uptake of sulphate from the cytosol into the chloroplast. This led them to suggest (Melis & Chen, 2005) that, given the well known effect of sulphate deprivation on photosynthesis (and hydrogenase activity) it might be possible, by mRNA interference technology, to select for mutants with photosynthesis depressed to the level of respiration. Such a mutant would be normally subjected to anaerobic conditions and might be expected to produce H_2 gas without the need for sulphate deprivation. Tolleter et al. (2011) developed a mutant form of *C. reinhardtii* with impaired cyclic electron flow (CEF) around photosystem I due to a defect in the protein regulating proton gradient (PGRL1). In the wild-type strain, the proton gradient generated by CEF provokes a strong inhibition of electron supply to the hydrogenase whereas in the mutant, this inhibition is released. This opens up the possibility of perhaps enhancing H_2 production by reprogramming cellular metabolism through regulating the trans-thylakoid proton gradient by monitoring expression of the *pgrl1* gene.

Chapter X

Microalgae and Bioremediation

Bioremediation processes involving the use of living organisms to treat various forms of aquatic pollution have been applied to degrade or eliminate industrial contaminants such as heavy metals, pesticides, oils, solvents and a range of other organic and inorganic chemicals. Microalgae are generally regarded as being less suitable for such processes because of the problems of harvesting and disposal of the resulting contaminated biomass and because they are often relatively intolerant of many of the most common pollutants. Toxic spills from a variety of industrial sources such as tailings impoundments have confirmed the susceptibility of microalgae generally to heavy metals and metalloids. There is evidence, however, that some microalgal species are able to develop resistance to heavy metal and other pollutants following prolonged exposure (Knauer et al. 1999; Costas et al. 2001; Lopez-Rodas et al. 2001) and there are many situations where microalgae have been successfully used in bioremediation projects. The potential contribution of microalgae to bioremediation systems, either through increasing the degredative capability of the microbial community as a whole or through direct transformation of specific pollutants, has been widely reported. Among the factors predicted to contribute to future improvements in microalgal-based treatment systems would be the application of genetic engineering methods to develop high product selectivity. Microalgae have also been extensively used in bioassays to assess the nutrient status of water bodies or as indicators of the presence of various toxic components. The varying sensitivity of the different species used in such bioassays requires careful evaluation in the development of standard tests for hazard evaluation schemes.

Microalgae as Indicators of Water Quality

The potential of microalgae as indicators of changes in water quality is based on their distinct ecological requirements and on their response to a variety of water quality parameters. Reid et al. (1995) in their review of the use of diatoms as indicators of past and present water quality in a range of aquatic environments, enumerated the various advantages that the diatoms possess in this regard. They include their sensitivity to both biotic and abiotic environmental changes and the predictability of that sensitivity. Another advantage that the diatoms have over other microalgal groups is their prevalence in a wide range of aquatic environments, their well-established taxonomy and physiology and, especially relevant in the interpretation of long-term changes in water quality, their preservation in the form of silica shells in, for example, lake sediments.

Techniques have been developed for using diatom communities to make quantitative estimates of past water chemistry. They are based on the classification of diatoms according to their tolerance of organic pollution (to yield a Diatom Assemblage Index of Pollution) which recognises categories ranging from oligo-saprobic (low tolerance) through meso- to poly-saprobic (high tolerance).

One such index of pollution is that developed by Watanabe et al. (1988) based on a study of over 500 aquatic sites in Japan. Some 454 diatom taxa were classified according to their pollution tolerance, with scores from 1 to 100 indicating the optimum conditions for each taxon within this tolerance range. The use of tolerance indices of dominant diatom taxa to assess water quality has been described by Round (1991, 1993) and applied to address problems of salinization and nutrient enrichment (e.g. Bennion 1994; Fritz et al. 1993; Yang & Dickman 1993; Wilson et al. 1994; Gell 1993). The value of diatoms as indicators of short-term, anthropogenic changes in water quality has been greatly enhanced by the development of more precise dating and sampling techniques such as those using Pb^{210} and Cs^{137} (Clark & Wasson 1986).

Various contaminants affect different species differently resulting in changes in the species composition of the diatom communities. By comparing diatom communities at different sites in the vicinity of Casey Research Station, Antarctica (where the benthic microflora consists almost entirely of diatoms – see Gilbert 1991), Cunningham et al. (2000) noted significant differences between contaminated and uncontaminated sites. To confirm that the differences between sites were a true indication of the effects of contamination, Cunningham et al. (2003) conducted follow-up manipulative field experiments in which uncontaminated, metal-contaminated and petroleum hydrocarbon-contaminated sediments were deployed for 11 weeks at different locations. They noted significant differences in the composition of diatom communities that developed in control (uncontaminated) and contaminated treatments.

Contamination by petroleum hydrocarbons had a stronger effect on the diatom community composition than was observed with metal contamination. Specifically, *Navicula directa* occurred at lower abundances in the hydrocarbon contaminated treatments than in corresponding control sites although another species *N. cancellata*, occurred at higher abundances in both hydrocarbon and metal exposed sites relative to the controls. The differences between species in their response to contamination and the consequent changes in benthic diatom communities were observable at contaminant concentrations comparable with those associated with moderately contaminated sites and within the time scale (11 weeks) of these tests.

There were also indications that communities in areas already stressed may be more susceptible to further imposed stresses (contamination or disturbance) than those from reference (uncontaminated) locations. It was concluded that with suitable care to distinguish between effects due to location and those due to contamination, careful analysis of benthic diatom samples might play a useful role in monitoring impacts caused by both metal and hydrocarbon contamination.

Microalgal-Based Bioassays for Assessing Heavy Metal Toxicity

The sensitivity of microalgae generally to the toxic effects of heavy metals, together with the ease with which most of them can be grown in pure culture, has led to their being extensively used in bioassays testing heavy metal contamination of various water bodies or industrial effluents. These bioassays measure the physiological and behavioural responses of the test microalga to short or long-term exposure. One such bioassay is the automatic ECOTOX system of Tahedl & Häder (1999) which uses the microalga *Euglena* as the test organism. Using a miniaturised microscope and infrared monitoring light source in association with a cuvette to which the exposed test suspension is supplied, they monitored effects on microalgal motility and on photosynthetic efficiency as measured by a non-invasive pulse amplitude modulated fluorimeter.

This device was used by Ahmed & Häder (2010) to monitor copper toxicity of *Euglena gracilis* as measured over short- (0-24h) and long-term (1-5 days) exposure. 50% inhibition index (EC_{50}) of motility (and the related parameters of orientation and upward swimming) was measured following exposure to copper ($CuCl_2 2H_2O$) within the range 300 – 360 μM. Similar levels of sensitivity were obtained following both short-term and long-term exposure. Although the main target for copper toxicity is photosynthetic efficiency and PSII electron transfer, bioassays based on motility measurements offer a rapid, sensitive and easily automated test of toxicity. It was noted that tests of long-term exposure (5 days) were, however, best observed as photosynthetic effects.

A range of freshwater microalgal species have also been extensively use for *in vitro* testing for the presence of toxic chemicals in waste waters. Such tests reqire careful standardisation both in the choice of the test microalga and in the conditions under which the test is carried out. It is known, for example, that the sensitivity of the microalgae, as measured by growth rate or final biomass after a standard time, is very much affected by the initial cell density of the culture (Franklin et al. 2002; Moreno-Garrido et al. 2000). This is explained by changes in the negatively charged binding sites at the microalgal cell surface for metal cations such as Zn^{2+} and Cd^{2+}.

Another important factor that should be considered is the previous history of the microalgal test organism in terms of exposure to heavy metal or other contamination. Thus, in their study of toxicity of cadmium and zinc, Monteiro et al. (2010) used as test organisms two microalgal species (*Scenedesmus obliquus* and *Desmodesmus pleiomorphus*) that had been previously isolated from a heavy metal polluted area in Northern Portugal. The effects on microalgal growth were determined after 96h exposure to several initial concentrations of the two metals. The results were expressed as the concentration of metal that inhibits growth (as biomass) of the microalgae by 50% (EC_{50}). For Cd the values were 0.58 and 1.92mgL^{-1} for *S. obliquus* and *D. pleiomorphus*, respectively, and for Zn the corresponding values were 16.99 and 4.87mgL^{-1}. These values indicated that the strains used in these tests were more tolerant of Zn than those described in the literarure, an observation that was attributed to the fact that this metal was an abundant contaminant of the original source site of the microalgal strains used. Of the two microalgal strains used, *S. obliquus* had a slightly higher capacity to remove Zn from the solution. Both microalgal strains were particularly sensitive to Cd, a metal known to bind to sulfhydryl groups important in regulating normal cell division (Visviki & Rachlin

1991) or in respiratory processes (Mohammed & Markert 2006). Moreover, much less Cd was taken up from the solution compared with Zn. On the basis of their findings in these tests Monteiro et al. (2010) recommend that such ecotoxicological tests should preferably be carried out using microalgal wild strains from already contaminated source sites so that any changes in the extent of contamination of the test sample will be directly reflected in the test results.

For assessing the degree of contamination of waste-water from industrial or other sources, Blanck & Wängberg (1988), proposed an ecotoxicological test, the "pollution-induced community tolerance" (PICT) which defines the community in terms of the sensitivity of its component species to a particular level of toxicity. It has been applied to assessing the effect of such toxic substances as arsenate and copper on freshwater phytoplankton communities (Wängberg 1995). Gustavson & Wängberg (1995) used enclosure experiments to investigated how a 20 day exposure to copper affected tolerance and succession of species in freshwater microalgal communities. Community tolerance of copper, measured as photosynthesis and as changes in species composition, was found to increase during the experiment. Communities exposed to 0.24μM Cu displayed increased tolerance after only 2 days; those exposed to a lower concentration of copper (0.016μM) required 12 days to show significant increased tolerance. The results also suggested common tolerance mechanisms for copper and zinc.

Because of its extensive use in a range of industrial manufacturing processes, chromium is a common pollutant of soils and waste-waters (Gutiérrez-Ruíz et al. 1990). The most stable and abundant forms of chromium as a pollutant are Cr(III) and CR(VI) (two oxidation states of chromium oxide), both well-known for their toxic and mutagenic effects on most organisms (Banerjee et al. 2008). Tests based on the response of various microalgae to chromium have shown that the metal exerts its toxic effect primarily through an effect on energy metabolism, specifically upon photosystem II activity (Rocchetta & Küpper 2009). In heterotrophic cultures of *Euglena gracilis*, the IC_{50} values for growth under exposure to CrVI and CrIII were 12 and >250μM, respectively (Jasso-Chávez et al. 2010). There was evidence that chromium cytotoxicity was due to altered gene expression and hence enzyme content, and induced oxidative stress.

Microalgal Bioassays for Other Pollutants

Microalgae have been extensively used to assess the effectiveness of purification processes used for treating waste from a variety of sources. Petukhov et al. (2000), used cultures of *Euglena gracilis* to test the toxicity of effluents contaminated by diesel fuel or kerosene. It was noted that there was a positive relationship between the concentration of pollutant and the length of the lag phase and subsequent growth rate of the microalgal cultures, with diesel fuel having a weaker toxic effect than kerosene on this microalga. The same microalga was also used to test the effectiveness of destructor strains of various bacteria in reducing toxicity of polluted effluents. The results confirmed that after prolonged exposure to effective destructor strains the effluent medium could support growth of the test microalga at levels close to that of the control level. *E. gracilis* was also used as the test organism in bioassessments of wastewater from an industrial estate in Peshawar, Pakistan (Azizullah et al.

2012). It was discovered that although a range of physicochemical tests showed that most of the water samples were within the acceptable range of national environmental quality standards for municipal and industrial effluents, they all adversely affected many of the parameters tested in the *E. gracilis* bioassay. Among the parameters adversely affected were motility, swimming velocity, cell shape and gravitactic orientation. Photosynthetic efficiency (measured using a pulse amplitude modulated fluorimeter) was the only parameter not affected.

A new class of polymeric materials (engineered nanoparticles, ENPs) is now commonly used for the preparation and targeted delivery of a range of biomedical products such as drugs or genetic components. Their increasing use raises concerns as to their possible toxic effects and on their potential environmental impacts (Colvin 2003). Some ENPs have been shown to have toxic effects on bacteria, on animal cells and on a range of microalgae (*Chlamydomons reinhardtii, Desmodesmus subspicatus, Pseudokirchneriella subcapitata* and *Thalassiosira weissflogii*) where the effect is largely attributed to metallic nanoparticles (e.g. Miao et al. 2009). Another group of ENPs, the dendrimers, branched organic polymers of repeating patterns emanating from a central core and an outer surface of terminal functional units, have been shown by Petit et al. (2010), to have significant toxic effects on the viability and photosynthesis of *Chlamydomonas reinhardtii*. Specifically, it was observed that different dendrimers had different effects including, in the case of some types of dendrimers, a stimulation of photosynthetic performance. It is likely that microalgal bioassays will, in future, come to play an important part in testing for any toxic effects arising from the increasing use of these novel therapeutic devices.

Microalgae in the Treatment of Heavy Metal Contamination

Industrial methods for removal of heavy metal cations from waste include precipitation with reagent precipitants, and stripping by ion exchange methods. The former method suffers from the fact that different metals precipitate at different pHs and from their reduced efficiency in the presence of complex-forming agents and under high salt concentrations. The latter method as well as being less efficient in solutions with high salt content, are also expensive. For these reason, biosorbents prepared from various micro-organisms are now being considered, including those occurring as natural biomats in which microalgae form an essential component (Zakharova & Parfenova 2007). Microalgal-based wastewater treatment systems are particularly suited to small-scale, rural applications where the high cost and technological complexity of conventional methods are often prohibitive. Stabilization ponds in which microalgae constitute a major component, widely used for nutrient removal (De la Noue et al. 1992; Olguin 2003), are increasingly used for the removal of heavy metals (Wilde & Benemann 1993). Oxidation ponds containing a mixture of microalgal species or of microalgae and other organisms are generally favoured. Kumar & Goyal (2010) treated domestic wastewater from a village in India (Punjab) using a microalgal and duckweed-based stabilisation pond. It was found that even during the winter months, when the pond was dominated by *Chlorella,* passage of the wastewater through the pond was effective in removal of Zn^{2+} and Pb^{2+} with 72% removal of the former and 73% removal of the latter. Despite the evidence for some Zinc toxicity of the *Chlorella* cells (at the higher concentrations of 20mg.L^{-1}) Pb produced no such toxic effects. Laboratory cultures of the marine diatom *Phaeodactylum tricornutum* react to exposure to heavy metals such as Cd, Pb, or Zn by synthesis of certain peptides with a high cysteine content, resembling the phytochelatins well known to be induced in higher plants exposed to heavy metals (di Toppi & Gabbrielli 1999). Such findings point to the probable mechanisms involved in microalgal-based cleansing systems and to their possible future improved efficiency through transgenic applications to improve collection and binding of metal contaminants (Bozarth et al. 2009).

One of the most hazardous of the heavy metals released into the environment as a waste product from a range of industrial and engineering processes is mercury. It is widely used in the manufacture of chlorine, caustic soda and a range of plastics, as well as in electrical components and batteries. Its toxicity is largely due to inactivation of enzymes through reaction of the mercuric ions with the enzymic sulfhydryl groups (-SH). Mercury is toxic in its inorganic forms and also in organic combination as monomethyl, dimethyl and phenyl forms. Methyl mercury, for example, has been reported to be a most hazardous form of the metal because of its high lipid solubility (Jensen & Jernelöv 1969). The use of microalgae as a more cost-effective alternative to chemical and electrochemical methods for removing mercury from contaminated soils and water bodies is based on the observed strong capacity of some species to bind the metal. Mercury accumulation within microalgal biomass (to levels up to several orders of magnitude higher than in the surrounding medium) has been described as being due to a combination of physical adsorption and metabolic processes (Manuel & Neyra-Tanabe 2012; De Filippis & Pallaghy 1976).

In a modification of the standard practice of cleansing waste water by passing it through concentrated microalgal cultures, Wilkinson et al. (1990) used layers of microalgal biomass immobilized on calcium alginate beads. They found that immobilized cells of *Chlorella emersonii* (10^5 – 10^6 cells per bead) were more effective than non-immobilized cells (at equivalent concentrations) in removing mercury from the supplied medium. They further found that loss of mercury from the system was largely due to volatilisation in the form of metallic mercury Hg^0 and that the volatilisation was most effective with alginate immobilisation matrices (compared with agarose or agar).

Effective microalgal treatment of mercury-contaminated effluents would clearly be enhanced in the presence of a mechanism for detoxification of the accumulated metal such as the NADPH-dependent mercuric reductase (Mer A) which mediates the reduction of Hg^{2+} to elemental Hg^0, and is known to be active in some bacteria (Barkay et al. 2003). Huang et al. (2006) were able to transform *Chlorella* cells by introducing the *merA* gene from *Bacillus megaterium* strain MB 1 into the microalgal cells. They used a bacterial isolate from Minamata Bay (Japan) sediment which was heavily polluted by mercury as a source of merA gene which, as an appropriately modified fusion vector DNA, was then introduced to *Chlorella* protoplasts (produced by cellulase treatment of the microalgal cells). Promoters were used to drive expression of the merA gene in *Chlorella* and the transformed protoplasts allowed to regenerate their cell walls before being plated out on an agar medium. Subsequent bioassays confirmed that successful introduction of merA gene had been achieved and that functional expression of merA in *Chlorella* had been accomplished with the merA transgenic strain showing a higher ability than the wild type for removal of Hg^{2+}. Moreover, the merA transgenic strains had higher growth rates and photosynthetic activity in the presence of toxic concentrations of Hg^{2+}.

Contamination of drinking water by arsenic of geological origin is a serious problem in India and Bangladesh. Maximum concentrations up to 900μgL^{-1} have been recorded in the Nadia district of West Bengal, far in excess of the 10μg L^{-1} limit recommended by World Health Organisation. Studies of tube wells in a heavily arsenic contaminated area of Barasat in West Bengal have shown the microalgal flora to be dominated by the cyanobacterium *Phormidium laminosum,* with an arsenic content of 276.9μg g^{-1}in its biopmass (compared with an As concentration of 244.13μg L^{-1} in a shallow well and 146.13μg L^{-1} in a deep well) (Bhattacharya & Pal 2011). There was evidence that the resistance of this cyanobacterium to As and its ability to accumulate As involves modulation of stress enzymes and stress-related compounds, pointing to its potential use in a water-treatment pond system. Imamul et al. (2007) reported effective bioremediation of arsenic toxicity by microalgae in rice crops. Microalgal species that have been trialled include a number of coccoid green algae as well as some cyanobacteria such as *Spirulina platensis*, effective in the biosorption of mercury from polluted water bodies (Parameswari et al. 2010).

Microalgae in Waste-Water Treatment

Intensive farming systems, including aquaculture, inevitably produce waste matter with high organic and high nutrient content, requiring the development of cost-effective and environmentally suitable effluent treatment methods. As an alternative to the traditional methods of activated sludge microbial processing or anaerobic digestion, microalgae have often been used to treat such effluents. The use of microalgae in waste water treatment processes offers the possibility of combining the cleansing effect of the microalgae with the production of microalgal biomass as a potential stock feed or for biofuel production (Benemann & Oswald 1996). This assumes that some of the extra costs incurred in running a microalgal–based cleansing system can be set against the value of the harvested biomass. Combined microalgal-bacterial systems have also been suggested (González et al. 2008) in which the microalgae would provide the O_2 to allow mineralisation of organic matter by aerobic bacteria and, at the same time, assimilate N and P into the biomass and oxidise NH_4^+. The microalgae would also utilize CO_2 released by bacterial respiration (Muñoz & Guieysse 2006).

Nutrient Removal

Microalgae are best suited to the tertiary stage of waste-water treatment where the main purpose is the removal of the high concentrations of nutrients (ammonium, nitrate and phosphate) that remain after primary and secondary treatment. Planktonic microalgae and cyanobacteria grown in artificial sewage-containing ponds have been widely used for their capacity to remove nutrients (Oswald 1988; Picot et al. 1991; Nurdegan & Oswald 1995, Hoffman 1998;. Garcia et al. 2000). Large-scale pond systems for combined wastewater treatment and biofuel production were successfully trialed by Craggs et al. (2012) at the Christchurch (N.Z) wastewater plant. The system consisted of 1.25ha earthen treatment ponds constructed from pre-existing conventional oxidation ponds. Concentrated microalgal

/bacterial biomass was harvested from the ponds to yield 1-2% organic solids for further dewatering. The naturally developed microalgae were dominated by species of *Micractinium* and *Desmodesmus*. Average treatment efficiencies over the first 15 months of operation were ~50-90% for BOD, ~65% for NH_4 –N and ~19% for removal of dissolved reactive P. Importantly, the ponds were self-sealing thus avoiding the expense of using plastic liners.

For maximum effectiveness, microalgal-based nutrient removal systems would have to include efficient harvesting and subsequent disposal of the microalgal biomass. Because of the high costs of harvesting by centrifugation or flocculation, a number of alternative methods have been employed, mostly involving immobilization of the microalgal cells through entrapment in carrageenan beads or gel beads (Chevalier & de la Noüe 1985). Periphyton biofilms are a common feature of both natural and artificial nutrient-rich streams (Sladeckova 1994, Przytocka-Jusiac 1984), and algal turf scrubbers have been extensively used to cleanse both freshwater and marine nutrient-rich effluents (Craggs et al. 2003, Kebede-Westhead et al. 2003, Hoffmann 1998). Their success has stimulated the development of a range of treatment systems using microalgae immobilised on different artificial substrates. Guzzon & Albertano (2009) studied the light-induced photosynthetic behaviour of phototrophic biofilms developing on artificial substrates in various sections of a waste water treatment plant. They found that the films of microalgae (including cyanobacteria) that developed were able to acclimatise to seasonal changes in irradiance and were able to maintain high rates of photosynthesis and biomass production under varying seasonal conditions confirming their potential as a year-round bioremediation tool in waste water treatment.

Immobolization techniques serve to avoid washout of the microalgal biomass while still maintaining efficient nutrient removal. Zamani et al. (2012) used screens composed of sodium alginate gel and found that orthophosphate removal of *ca.* 70% was obtained with *Scenedesmus rubescence* and *Chroococcus disperses* over 12 days of operation. Shi et al. (2007) used a twin-layer system in which the microalgae (*Chlorella vulgaris* and *Scenedesmus rubescens*) were immobilized by self-adhesion on a wet, microporous ultrathin substrate layer composed of reinforced nitrocellulose on either side of the middle source layer (composed of macroporous glass fibre material) through which the aqueous waste water medium is pumped (Figure 14). The system allows diffusion of nutrients from the source layer to the microalgal cells which remained attached to the outside layer. In a trial using municipal wastewater previously subjected to activated sludge treatment, the twin-layer system was very effective (with both microalgal species) in removing nitrate. A second trial using synthetic secondary wastewater showed the system to be effective in removing phosphate, ammonium and nitrate to less than 10% of their initial concentration over a period of days.

Systems based on the use of microalgae in combination with other microorganisms were used by Davis et al. (1990) and by Sabater et al. (2002). González et al. (2008) compared the effectiveness of *Chlorella sorokiniana* on its own and in combination with a bacterial component, for treating the effluent from piggery wastewaters. They found that while the microalgal- bacterial systems worked well with slurry that had been diluted four or eight times, the biodegradation process was much reduced in undiluted or two times diluted wastewater. Improved performance was achieved, even with up to two times diluted slurries (containing up to 1,180 mg NH_4^+ -N l^{-1}) when the pH was adjusted daily to 7 in enclosed microalgal-bacterial tests. High pH levels and high NH_4^+ concentrations were thus identified as the major factors influencing performance of the biodegradation process.

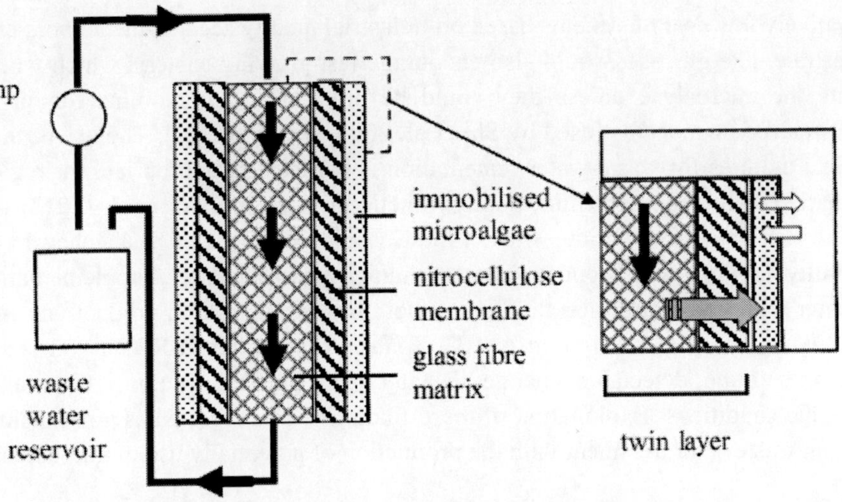

Figure 14. The twin-layer wastewater treatment system showing the outer microalgal layers immobilized on the reinforced nitrocellulose substrate layer and the central loosely- packed glass fibre layer through which the source wastewater is pumped (solid arrow). The large broken arrow depicts the diffusion of dissolved nutrients to the microalgal layer; the two-way open arrows depict diffusion of gases (CO_2 and O_2) into and out of the microalgal layer. (Redrawn, with permission, from Shi et al. 2007).

The main NH_4^+-N removal process was via assimilation into microalgal-bacterial biomass and nitrification to NO_2^- while a large fraction of the removal of soluble organic carbon occurred through biodegradation of volatile fatty acids (mainly acetic and propionic acids).

Effective nutrient removal, through incorporation into microalgal biomass, requires the microalgae to be actively growing which is difficult to sustain by photosynthesis alone, given the invariably dense cultures that build up in large bioreactors. However, the ability of many microalgae to undergo heterotrophic growth using organic compounds as a source of carbon and energy offers a promising solution. Unfortunately, although domestic and municipal waste-water that has passed through the microbial secondary treatment stage invariably contains large amounts of dissolved organic matter, much of it is in the form of large molecules not available for metabolism by microalgae. Recognising this, Perez-Garcia et al. (2011) tested the effect of supplementing municipal waste-water (or the synthetic equivalent) with a range of organic supplements to stimulate heterotrophic growth of *Chlorella vulgaris.* They found that the best carbon sources for heterotrophic growth were Na-acetate and D-glucose which provided the highest growth rates and the greatest removal of ammonium. The optimum concentration for Na-acetate was 0.12M which was completely consumed over a 10 day period of incubation and resulted in a microalgal population one level of magnitude higher than could be achieved under autotrophic conditions.

Na-acetate was found in these tests to be slightly superior to D-glucose both for supporting heterotrophic growth of *C. vulgaris* and for removal of ammonium. The effectiveness of these two carbon sources was related to their ease of uptake into the microalgal cells (a monocarboxylic/proton symport system in the case of acetate and an active hexose/proton symport system in the case of glucose) and their subsequent assimilation within the cell (via acetyl-CoA synthase and the glyoxylate and TCA cycles for acetate and through the pentose phosphate pathway and TCA cycle for glucose). In noting the comparatively low cost of systems based on industrial quality acetate, the authors caution that this substrate, like glucose, would also encourage fast growing bacteria which would tend to dominate the microalgae unless they could be protected by immobilization in a suitable polymer matrix (such as that used by Shi et al. 2007, see Figure 14). The use of mixotrophic microalgal cultures for waste-water remediation is particularly appropriate for regions where the climatic conditions might limit photosynthetic production. Park et al. (2012) isolated 14 strains of micralgae from various sites in the Canadian province of Saskatchewan and tested their ability to grow in secondary-treated municipal wastewater. Supplementation of the wastewater with glycerol or acetate boosted both biomass and lipid productivity in the three strains *Chlamydomonas debaryana*-AMB1, *Chlorella sorokiniana*-RBD8 and *Micractinium* sp.-RB1b, with no detectable change in fatty acid profiles resulting from exposure to mixotrophic conditions. Exploitation of these findings was suggested as a promising strategy for linking wastewater treatment with the production of potentially useful microalgal biomass.

Biotransformation of Organic Compounds by Microalgae

The extensive use of a range of pesticides and herbicides in agriculture, horticulture and aquaculture together with the increasing use of a range of organic solvents in the chemical processing industry raises a number of problems in relation to both human health and wider environmental concerns. The most commonly used solvents are volatile compounds such as methanol, ethanol, isopropanol, acetone, acetonitrile and dimethylformamide, compounds widely used in multiphase bioprocess operations including liquid-liquid extraction of a range of products. The use of these solvents requires careful monitoring of the effluents and industrial wastewaters to guard against spillages and to measure possible harmful effects on the environment. The sensitivity of microalgae to the toxic effects of such compounds has been widely reported and applied as biosensors to test for the level of contamination.

The green microalga *Pseudokirchneriella subcapitata* has been recommended by the OECD as a suitable standard biological indicator for such tests. It was used by Cho et al. (2009) to assess the toxicity of a range of commonly used organic solvents as measured by effects on photosynthetic activity. The inhibitory effects of the tested solvents were in the order: isopropanol>acetone>acetonitrile>ethanol>dimethylformamide>methanol, with EC_{50} values ranging from 589 to 2,570 mM. The effects measured were directly related to the hydrophobicity of the tested solvents.

There is, however, increasing evidence that some microalgae are able to act on a range of organic compounds such as dyes, pesticides and insecticides, crude oil and a range of xenobiotics, reducing their harmful effects on the environment and producing non-toxic products for possible reuse such as in the production of single-cell protein (Queiroz et al. 2007). Some examples of the involvement of microalgae in the biodegradation of different hazardous pollutants have been cited by Ghasemi et al. (2011). They include the degradation of phenolic compounds by the chrysophyte *Ochromonas danica* through oxidative hydroxylation of the phenol ring (Semple et al. 1999). Other microalgae shown to be capable of degrading phenol are the cyanobacteria *Synechococcus* PCC 7002 (Wurster et al. 2003) and *Phormidium valderianum* (Shashirekha et al. 1997). Nitrophenols and their derivatives, common cantaminants from the industrial production of dyes and pharmaceuticals and from the agricultural use of pesticides and herbicides have, in pilot trials, been successfully degraded by a number of microalgal species including *Chlamydomonas reinhardtii*, and two species of *Anabaena* (Hirooka et al.2006). With these contaminants, as with many others, best results are obtained with mixed microalgal-bacterial culture systems (Borde et al. 2003). In some tests, such as those carried out by Semple et al. (1999) and Lei et al. (2002), quite complex, high molecular weight polycyclic pollutants (such as pyrene) have been shown to be susceptible to attack by microalgal cultures, partly through sorbtion effects related to the cellwalls and partly through intracellular chemical reactions. Other organic pollutants that have been successfully treated by microalgal-based systems include chlorinated organochemicals, organophosphate pesticides, steroids, terpenes and fuel oils (see the comprehensive review by Ghasemi et al. (2011). Cleansing systems based on microalgae offer many advantages over those using other micro-organisms, in particular, their suitability for largescale applications by the use of stabilizing ponds or other bulk culture methods.

Some microalgal species have been tested for their effectiveness in treatments aimed at alleviating the harmful effects of dioxin and dioxin-related compounds (polychlorinated-*p*-dioxin, PCDD, and polychlorinated dibenzofuran, PCDF) upon human and animal health. The harmful effects of such compounds (including cancer promotion) are exacerbated by their lipophilic nature and hence their tendency to accumulate in the fatty tissues of animal and fish products. Among the most effective treatments is the capture of the ingested dioxin product in the digestive tract, so preventing its absorption. The inclusion of *Chlorella* in the diet of experimental rats has been reported to accelerate dioxin excretion (Morita et al. 1999) suggesting a potential contribution from *Chlorella* (and perhaps other microalgae) to the on-going search for so-called interceptor molecules effective in protecting against these most highly toxic of synthetic molecules.

One of the world's most toxic and persistent of organochlorine peticides is the compound γ-hexachlorocyclohexane (lindane) which, despite its being prohibited in many countries, remains as a dangerous contaminant in many agricultural soils. Among the microorganisms currently in use for the bioremediation of lindane-contaminated effluents are certain cyanobacterial species (El-Bestawy et al. 2007). *Anabaena azotica* isolated from Chinese paddy soils has been found to tolerate exposure to lindane at an initial concentration of 0.2 mg L^{-1} and to be effective in degrading the pesticide with an efficiency of 48.8% after 5 days exposure (Zhang et al. 2012).

The production of microalgal biomass as the starter material for a range of biosynthetic processes is potentially most economically achieved when it is combined with a microalgal-based waste-water treatment system. This would require, of course, for the waste-water-produced microalgae to be of a composition suitable for the biosynthetic process concerned. Mulbry et al. (2008) investigated the composition of mass cultured microalgae resulting from the treatment of dairy and swine manure effluents. They examined specifically how the fatty acid content of the microalgal biomass, important for its possible use as a substrate for biofuel production, was affected by changes in the nature of the manure effluent, its loading rate and how it was affected under conditions of supplemental CO_2 supply. In their tests, the microalgal component (dominated by the filamentous green microalga *Rhizoclonium* sp.) was immobilised on to algal turf "scrubber" (ATS) units over which the effluent (either raw or anaerobically digested) was passed at loading rates ranging from 0.2 to 2.3g total N m^{-2} day^{-1}. It was observed that the fatty acid content of the ATS biomass (0.6 to 1.5% of DW) was at the very low end of values previously reported for nitrogen sufficient cultures of chlorophytes such as *Chlorella* and *Scenedesmus* (Piorreck & Pohl 1984; Shifrin & Chisholm 1981). The fatty acid composition was remarkably consistent across the treatments tested with >90% occurring as long-chain (14-18C) polyunsaturated molecular structures.

Extrapolations from pilot-scale ATS cleansing systems having average productivity values of 10g $DW.m^{-2}.day^{-1}$ for 270days of the year (Mulbry et al. 2005; Pizarro et al. 2006), have suggested a yearly productivity value 27,000 kg DW algal biomass. ha^{-1} for a loading rate of 2,700 kg total nitrogen, 400 kg total phosphorus ha^{-1}. Mulbry et al. (2008) estimated the operational costs for their "scrubber" system as \$1.30 (US) kg^{-1} algal biomass produced, without putting any value on the algal biomass nor on the cleansing process itself. This, they conclude, compares very favourably with the cost of \$32 US kg^{-1} for microalgal biomass produced in outdoor photobioreactors (Molina Grima et al. 2003). As feed stock for fatty acid production, however, the cost was estimated as an unacceptably high \$130US kg^{-1}fatty acid. Viewed purely as a waste-water treatment facility, the costs of an ATS system (estimated by Mulbry et al. as \$11US kg^{1}N) compares favourably with other such agricultural management practices, especially when used in areas of particular sensitivity such as the Chesapeake Bay area (see Chesapeake Bay Foundation Report, 2004).

Oil-polluted sediments often become colonised by dense vertically structured mats having various cyanobacterial species as a major component (van Gemerden 1993). Such microbial mat communities have been shown to have the potential to degrade various oil compounds, mainly through the ability of the bacterial component to break down various aliphatic and aromatic hydrocarbons (Abed & Köster 2005). A possible supporting role for the cyanobacterial component (dominated by *Halomicronema excentricum* and various species of *Phormidium*) has been suggested by Abed & Golubic 2009).

Induced Resistance of Microalgae to Pollutants

Some of the most serious pollutants of aquatic environments occur in the form of pesticide run-off from agriculture, antibiotics and hormones from medical effluents and a range of other miscellaneous chemical contaminants from industrial and military activities. The extent to which microalgae can adapt to exposure to such pollutants rests largely on genetic mutation as has been demonstrated for many microalgal species with respect, for example, to herbicide resistance (Boura-Halfon et al. 1997: Costas et al. 2001). Lopez-Rodas et al. (2001) studied adaptation of some microalgal species to different pollutants including the antibiotic erythromycin and the nitroaromatic explosive compound 2,4,6-trinitrotoluene (TNT). They found that when such pollutants (at concentrations of up to 10ppm for the former and up to 28ppm for the latter) were added to dense microalgal cultures of, for example *Pseudanabaena planctonica* (cyanobacterium) or *Scenedesmus* sp. (Chlorophyceae), the culture density was reduced after a few days due to death of sensitive cells. After further incubation (for several days), however, there was sometimes an increase of cell density due, it was shown, to the growth of mutants resistant to the contaminants. Further analysis confirmed that pollutant-resistant cells arose randomly by rare spontaneous mutation during cell replication prior to addition of the pollutant (pre-selective mutation). The rate of spontaneous mutation from sensitivity to resistance ranged from $2.0x10^{-6}$ mutants per cell division (*P.*

planctonica for erythromycin) to 8.2x10^{-6} mutants per cell division (*Scenedesmus* sp. for TNT). Since pollutant-resistant mutants have a diminished fitness, the resistant variants are maintained, in the absence of pollutants, by the balance between new resistant cells arising from spontaneous mutation and resistant cells eliminated by natural selection. Such rare, spontaneous pre-selective mutations, it was concluded, ensures survival of a microalgal population exposed to a polluted environment.

The ability of microalgae to adapt following exposure to a toxic spill of acidic heavy metals was investigated by Baos et al. (2002) using mud/soil samples taken from an area affected by an environmental disaster at the Aznalcóllar pyrites processing facility in Spain (April 1998). Samples containing heavy metals at different concentrations were inoculated with mid-log phase cultures of *Scenedesmus intermedius* and the response expressed in terms of the carrying capacity of a particular assay and as the mutation rate under those particular assay condition. It was thus possibel to distinguish between resistant cells arising by rare spontaneous mutations occurring randomly during replication and those arising through specifically acquired adaptation in response to environmental selection. The results indicated that resistant cells arose primarily by the former mechanism and that such rare, pre-adaptive mutations were enough to ensure survival of the microalgal population in a contaminated environment provided the population is large enough. Copper-tolerant mutants of *Chlamydomonas reinhardtii* have recently been induced by genetic insertional methods and have proved useful in studies of the mechanism of copper toxicity under both autotrophic and mixotrophic conditions (Sabatini et al. 2011).

Chapter XI

Microalgae and Carbon Capture

In view of the high cost of chemical or physical methods for absorbing gaseous CO_2, it is likely that photosynthesis will remain for some time the most feasible method for restraining the continuing build-up of CO_2 resulting from fossil fuel consumption. Of all the photosynthetic organisms, microalgae would appear to be well suited for this purpose compared with terrestrial plants which, anyway, have other more valuable specialist uses as crop or timber sources. Although prototype microalgal-based carbon capture plants have been extensively trialled as demonstration projects, it is likely that, for the immediate future, the role of microalgae in carbon capture will be largely restricted to natural, mostly marine, populations. In an assessment of the potential of marine photosynthesis as a sink for anthropogenic CO_2 emissions, Nellemann et al. (2009) estimated that marine primary producers contribute at least 50% of the world's carbon fixation and may account for as much as 71% of all carbon storage in oceanic sediments. Over the vast areas of the world's oceans, such sequestration processes are due primarily to the microalgal component although at inshore and estuarine regions there would be significant contributions from other primary producers such as mangroves, seagrasses and macroalgae (Chung et al. 2011).

Sequestration of Carbon in Oceanic Depths

Concerns over the observed global increase in the concentrations of atmospheric CO_2 has focussed attention on the potential of oceanic biota to sustain flux of biogenic carbon into deep waters so mitigating the effects of climate change. Legendre & Fèvre (1992) defined biogenic carbon (BC) pools in the ocean in terms of turnover times, defined as the time (τ) between photosynthetic uptake of carbon by phytoplankton and return of this carbon as CO_2 to the surface waters or the atmosphere. Short-lived organic pools (τ=3-4days) comprise mainly microbial (including microalgal) components of the food web. Long-lived organic C pools include renewable marine resources (eg. fish and marine mammals) and heterotrophic bacteria that break down organic matter derived from large heterotrophs. The third pool comprises sequestered biogenic carbon and includes organic remains buried in sediments, continental limestone, refractory dissolved organic matter and dissolved CO_2 in deep waters. Figure 15 shows a conceptual model for food-web mediated C flux from phytoplankton primary production through the various pools.

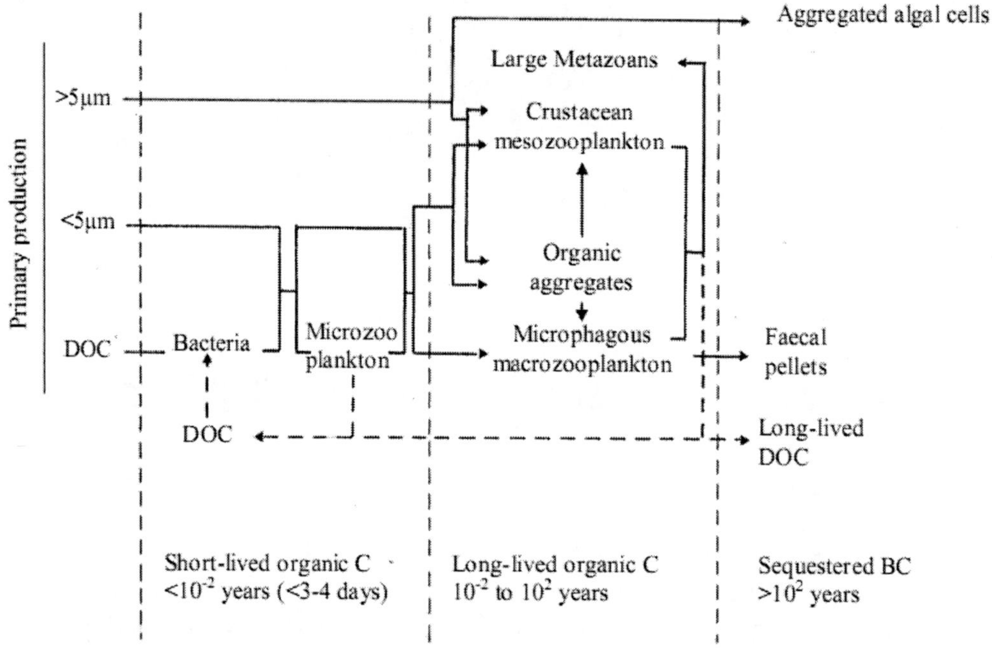

Figure 15. Conceptual food-web mediated C flux in oceanic waters. Solid arrows show major flows of biogenic carbon (BC) in the euphotic zone from three size classes of net primary production – larger microalgae (>5µm), smaller microalgae and picophytoplankton (<5µm) and dissolved organic carbon (DOC) to three carbon pools (short-lived, long-lived and sequestered). Dashed arrows show food-web recycling of DOC, from consumers to heterotrophic bacteria (the microbial loop) and carbon sequestration as DOC. (Redrawn, with permission, from Legendre & Michaud, 1998).

The critical component of this model is the speed with which BC sinks out of the upper water column (i.e. becomes sequestered). Legendre & Michaud (1998) developed equations to quantify the depth from which BC cannot readily return to the surface waters or the atmosphere. They defined the main mechanism for sedimentation of organic particles as their aggregation into larger, faster sinking particles. They recognised that living, healthy organisms generally do not sink or, if they do, they sink only slowly. Senescent or dead organisms, on the other hand, tend to aggregate in large organic particles, so increasing their sinking velocity. Sinking organic particles include marine snow (aggregates of various particles, including phytoplankton) and faecal pellets.

The microalgal component through its position at the base of pelagic food webs represents a major influence on biogenic carbon flux, determining flux towards pools of long-lived organic C and sequestered BC. Whether the microalgal biomass sinks, is preyed upon, or undergoes lysis will determine the amount of carbon that reaches the sediment and remains buried. Zooplankton grazing will channel microalgal biomass to higher trophic levels whereas release of microalgal constituents by lysis or microbial action will affect the standing stock of dissolved (DOC) or particulate (POC) organic carbon, forcing the food web toward more regenerative pathways. Viral-induced and other forms of microalgal lysis decrease the efficiency of carbon-transfer to higher trophic levels (Fuhrman 1999; Suttle 2007). Viral lysis, in particular, is thought to divert a significant fraction of the photosynthesized carbon pool into organic debris in both oligotrophic and eutrophic marine ecosystems (Danovaro et al. 2010). Further details of the involvement of different microalgal species in the transportation of photosynthetically fixed carbon to oceanic depths and to the sediment have been described in Chapter III.

Separation of Greenhouse Gases from Industrial Effluents

The apparent effectiveness of marine phytoplankton in triggering the transportation of carbon from the atmosphere to oceanic depths has led to speculation as to whether microalgal photosynthesis might also have a role to play in removing CO_2 from various other sources such as industrial emissions. One of the cost advantages of using microalgal cultures for taking up CO_2 compared, for example, with chemical or physical methods, is that the resulting microalgal biomass could itself constitute a useable resource such as feed stock or biofuel or, alternatively, would contain a high percentage of carbon in a form (i.e. mineralised carbon) suitable for sequestration. To make a significant contribution to reducing atmospheric concentrations of CO_2 or other "greenhouse gasses", however, the microalgal biofuel would have to replace fuels that would otherwise come from non-renewable sources (i.e. fossil fuels) and would have to be less polluting in terms of the yield of undesirable pollutants, such as sulphur compounds. Similarly, the microalgal biomass would have to be used for the production of other useful compounds like oils or plastics that normally come from non-renewable sources. Its use as stock fodder or aquaculture feed would be most effective if, through manipulation of the culture conditions, it resulted in a product giving a lower yield of other "greenhouse gases" such as methane from cattle. The flow of carbon through such a functioning microalgal-based carbon-capture process is outlined in Figure 16, in which CO_2 is diverted from an industrial effluent source to a photo-bioreactor in which microalgae use solar energy to produce biomass for further processing or sequestration.

Some of the advantages of such a system have been listed by Olaizola (2003) as

1. Flue gas containing CO_2 could be fed directly to the microalgal culture
2. Some other flue gas components such as NO_x or SO_x may also support microalgal growth
3. The operating costs of the process could be partially met from the sale of high-value commercial products
4. Under ideal conditions, the process would function with minimal negative impacts on the environment.

In addition to CO_2, flue gases also contain other greenhouse gases (GHG), most notably various NO_x gases and, depending on the conditions under which the microalgal cultures are maintained, there may also be significant emission of other GHGs such as CH_4 and N_2O. It has been shown, for example, that microalgae are generally able to tolerate high concentrations of CO_2 (Hanagata et al. 1992) as well as being able to grow in the presence of oxygenated compounds of S and N (Negoro et al. 1991;Lee et al. 2002), common constituents of flue gases. Olaizola et al. (2003) tested the ability of cultures of the green microalga *Haematococcus pluvialis* to capture CO_2 from coal and propane combustion gases (and other simulated gas mixtures) fed to photobioreactors maintained at different pHs.

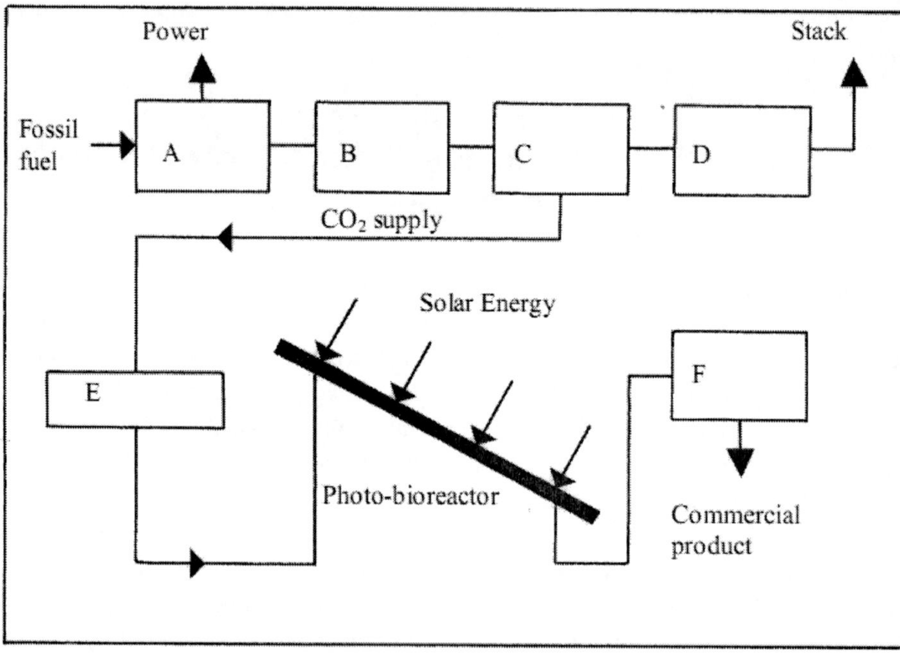

Figure 16. A summary of the major components of a microalgal-based carbon capture scheme for treatment of industrial flue-gas. A – Combustion system, B – Pre-process, C – CO_2 removal, D – Post-process, E – Pre-process, F – Microalgal harvesting, commercial products, carbon sequestration. (Redrawn, with permission, from Olaizola et al. 2003).

They found that, under optimal conditions, the microalgal culture was able to capture up to 70% of the available CO_2 when the culture was maintained at pH 7.5, with reduced capture of CO_2 at higher pHs. Harter et al. (2012) assessed the flow of carbon and nitrogen through cultures of *Dunaliella salina* supplied with biodiesel flue gas and observed that despite low sequestration of CO_2 and moderate retention of NO_x–N, the emission of CH_4 and N_2O.could result in an overall negative GHG balance for the system as a whole. The fact that the emissions were lower in pre-sterilized cultures indicated that the major source of the GHG emissions was anaerobic microbial processes, suggesting that sequestration systems using open ponds, with low mixing, anaerobic conditions and contaminating organisms would not be very effective.

One of the major sources of CO_2 emissions from industrial souces is that associated with cement production, where the gas is produced partly during the provision of energy to support various energy requiring processes and partly from the chemical calcination of limestone (Metz et al. 2007). Borkenstein et al. (2011) fed flue gas from a cement plant by airlift to a culture of *Chlorella emersonii* in an illuminated 5.5L glass reaction vessel, with the rate of supply of the gas controlled by the pH of the culture medium. The flue gas at this plant (in Retznei, Austria) came from the heating of limestone using stone coal and various surrogate fuels such a tyres, plastics and meat and bone meals. The pH-stat photobioreactor yielded, over a period of 30 days, a total of 2g L^{-1} microalgal dry mass (at a growth rate of 0.13 day^{-1} and a CO_2 fixation rate of 3.25g L^{-1}). No accumulation of flue-gas residues were detected in the culture medium but the microalgal biomass contained lead at three times the concentration in a control bioreactor fed with pure CO_2. The latter finding would preclude use of the microalgal biomass from this plant as animal feed but not for other processes such as biodeisel production.

A number of studies have used larger reactor vessels of different configurations to expose different microalgal cultures to flue gases from various sources such as kerosene- and solid fuel-fed thermal power plants (Maeda et al. 1995) and waste incinerators (Douskova et al. 2009) without significant negative effects upon microalgal growth compared with gas mixtures made up with pure CO_2. Doucha et al. (2005) successfully combined several 400L units to grow *Chlorella* in cultures of total capacity close to $500m^3$. Search for new microalgal strains effective in removing CO_2 from flu gas emissions from a sewage treatment plant in Tokyo, Japan (Satoh et al. 2009) revealed that the most effective was identified as belonging to the genus *Parachlorella,* designated *P.*sp.*binos*. A ten year Japanese study (RITE biological CO_2 fixation project, reported by Kurano, 2009) achieved CO_2 fixation rates of 50g $CO_2\ m^{-2}\ d^{-1}$ in a photobioreactor under outdoor conditions on land (1 km^2) adjacent to a 1000 MW power plant. The microalgal biomass product was, moreover, suitable as poultry feed although at a cost far in excess of that achieved by using imported feed such as corn. Among other problems is the reported observation that for at least some microalgal species (e.g. *Chlorococcum littorale*) sudden changes in CO_2 concentrations induced undesired physiological responses.

The existence in many microalgal species of carbon concentrating mechanisms (CCM) to enhance the supply of CO_2 for photosynthetic carbon fixation (Suzuki et al. 1994) has been cited as an area for possible improved carbon capture performance. Studies with the red microalga *Porphyridium purpureum,* for example, have shown that its CCM contains a novel catalytic site for CO_2 hydration (Mitsuhashi et al. 2000). Another red microalga (*Cyanidium caldarium*) has been reported to survive and grow at very high CO_2 concentrations (Seckbach et al. 1970) as has the green microalga *Chlorococcum littorale* which grows well when bubbled with air enriched with 60% CO_2 (Kodama et al. 1993), well in excess of the 10-20% CO_2 common in most industrial flue gases. One of the observed immediate effects of exposure of cultures of *C. littorale* to high CO_2 concentrations was the few days lag before a more active growth rate was established (Pesheva et al. 1994). PSII activity was suppressed during the lag period (perhaps due to intracellular acidification) and PSI activity enhanced. Kurano & Miyachi (2004) have suggested that acclimation of *C. littorale* to high CO_2 (through a large increase in the PSI/PSII ratio) involves enhanced H^+-translocating ATPase activity to maintain cytoplasmic and chloroplastic pH values suitable for growth. Using a flat-plate photobioreactor, Kurano & Miyachi obtained growth rates of *C. littorale* corresponding to a CO_2 uptake of 16.7 g over a 24h period. Coccolithophorid microalgae, because of their extracellular scales of $CaCO_3$, might be expected to be particularly suitable for CO_2 capture systems. Moheimani & Borowitzka (2007) grew the haptophyte *Pleurochrysis carterae* in semi-contiuous culture in outdoor raceway ponds and found that by adjusting raceway depth to suit the prevailing environmental conditions (and by countering the early morning inhibitory effects of high O_2 concentrations) they could maximise productivity and hence carbon capture potential.

Chapter XII

Conclusion

It has been suggested that the three major challenges facing humankind in the 21st century relate to food, energy and the environment, including climate change and environmental degredation due to pollution and habitat loss. As this review has shown, microalgae have the potential to contribute, to varying extents, to each of these areas of sustainability. In the area of fisheries and food production, the contribution of microalgae, already considerable, is predicted to continue to grow into the future. In some other areas such as bioremediation and the production of high value compounds, current technology is restricted to a relatively small number of applications and limited by cost considerations and low competitiveness with alternative technologies. Finally, in the area of biofuel (including hydrogen) production, the technology has scarcely progressed beyond the pilot plant stage, and any future large-scale applications will depend on possible technological advances and a realistic assessment of the balance between economic and environmental considerations.

Implicit in these developments will be greater understanding of microalgal functions. This will require adaptation of new technologies, some already available from developments in other areas of biological research, others yet to emerge from expansion of current research on microalgae, including work on species outside the small group currently dominating research in this area. It is instructive to note that, as a proportion of total biological research effort (in terms of funding and instrumentation) that devoted to microalgal research and technological development seems very small and certainly out of proportion with the current and potential future importance of microalgae to mankind. A comprehensive review of the status of microalgal mass culture as a source of pharmaceuticals and other biologically active compounds was published by Borowitzka in 1995. Since then, as this review has shown, there has been considerable progress and it is anticipated that the next few years will see greater recognition of the potential of microalgal biomass to contribute to a range of commercial applications.

It is predicted that future advances will proceed along two fronts; one based on the application of advanced biotechnology including molecular biology and bioengineering, the other on expansion of low-cost, community-based practices already in wide and successful use in many parts of the world. An important part of the on-going program of biotechnological development will be the careful selection of microalgal species most suited to a particular application. This will require analysis of the existing and ever expanding data base covering various aspects of the growth and functioning of a wide range of microalgal species. Such was the approach adopted, for example, by Picardo et al. (2012) who developed a computer-aided method for selecting microalgal species appropriate for three application categories namely biomass heating value, syngas production from biomass and bio-H_2 production. For lipid productivity and energy applications, their analysis confirmed two species, *Botryococcus braunii* and *Isochrysis galbana,* as the favoured species. Other applications such as the production of specific high-value compounds would, they point out, require expansion of the data bank and modification of the screening methodology used by them.

Research over recent decades has brought an increasing appreciation of the relative importance in microalgal biology of those species occurring at the lower end of the phytoplankton size range, the picophytoplankton (Chisholm et al. 1988). Their major role in the productivity of oceanic waters and to biogeochemical cycles of marine and freshwater ecosystems suggests a potential for a significant contribution to the various practical applications described in this review. The application of techniques of genetic molecular analysis has already contributed substantially to a better understanding of the taxonomy of this important group of microalgae. Some taxa previously provisionally identified as *Chlorella*-like and *Nannochloris*-like, have now been revised. The latter, for example, now constitutes the new genus *Picochlorum* (Henley et al. 2004). More recently another new picoeukaryotic microalga *Chloroparva pannonica* has been described and classified as belonging to the Trebouxiophyceae (the group that also contains *Chlorella* and *Nannochloris*) (Somogyi et al. 2011). Development of genomics to sequence the genome of the wide diversity of microalgae – especially at the picoplankton size range below 2μm – has shown that eukaryotic microalgal diversity is as wide as that of prokaryotes (Vaulot et al. 2008). PCR-based molecular studies at Lake Balaton (Hungary) have shown that a single vertical sample of the water column contains a large number of picocyanobacterial genotypes, including some previously unidentified groups, as well as the picoeukaryotes which could reach extremely high abundances in winter (Felföldi et al. 2011).

The globally abundant prokaryote *Prochlorococcus*, with its unique divinyl chlorophyll pigment, is now recognised as having a streamlined genome of some 2,000 genes (Rocap et al. 2003), compared with the *ca.*10,000 genes of the eukaryotic microalgae. Its small cell size and its high surface to volume ratio are a key to its competitiveness under conditions of limiting light and low nutrients and to its role as part of a tightly coupled loop of recycling nitrogen. The discovery by Martiny et al. (2009) that some strains of *Prochlorococcus* may contain the nitrate reductase gene, hitherto considered to be absent from this prochlorophyte, alters our perceptions of its contribution to marine food webs, especially in relation to the exporting of carbon from the upper ocean either through direct sinking of the picophytoplankton cells or following packaging in dense zooplankton faecal pellets.

Even within the well-researched genus *Chlorella,* recent genomic analyses have shed new light on various aspects of its biology. The genome of *Chlorella variabilis*, for example, is now known to include protein-encoding genes that encode for meiosis-specific proteins, despite the apparent lack of a haploid phase in that genus (Blanc et al. 2010). Similarly, other *Chlorella* strains, despite their non-flagellate nature, have been shown to produce homologues of *Chlamydomonas* flagellar proteins. The presence in the *Chlorella* cell wall of glucosamine polymers such as chitin and chitosan, rather than the cellulose and hemicelluloses typical of plant cell walls, has been interpreted to suggest that components of the chitin metabolism of *Chlorella* might have been acquired via lateral gene transfer from viruses. This suggests that the large DNA viruses predominantly associated with microalgae and marine protists may have played a greater role than previously recognised in the evolution of their hosts. The observed infection of some microalgae by large double-stranded DNA viruses may offer interesting insights into the significance of microalgal-viral interactions generally (Eckard, 2010) with possible implications for their biotechnological exploitation. The numerous examples of naturally occurring stable associations between various microalgal endosymbionts and a range of hosts offer interesting model systems for investigating further genomic interactions involving microalgae. One of the best known of such associations is that between microalgal zooxanthellae and their coral host cells. Recent reports on the nature of the metabolic interactions between symbiotic dinoflagellate toxins and host calcification (reviewed by McConnaughey 2012) may yet prove crucial to our understanding of the factors affecting growth and survival of the world's coral reefs and their adaptation to climate change.

The potential of genetic engineering to contribute to increasing the yield of various desired compounds by microalgae received a boost with the heterologous expression of a functional glucose transporter in the diatom *Phaeodactylum*. This endowed a hitherto obligate photoautotroph with the capacity to grow on glucose in the dark (Zaslavskaia et al. 2001) so opening up the possibility of substantially improved yield of, for example, lipid for biofuel production. Transformation protocols have now been developed for several diatom species, including the already sequenced *Phaeodactylum tricornutum* and *Thalassiosira pseudonana*, allowing random but stable integration of foreign DNA into the cells, leading to their possible use for directed synthesis of valuable products.

The unique ultrastructural patterning of the silica-based shells of diatoms is now recognised for its potential in a range of biotechnological applications. The extensive use of diatoms in filtration and adsorbtion processes has been extended to include DNA purification (Gilmore et al. 1993). They also have a role in the adsorbtion of heavy metals and other contaminants from industrial wastewater effluents (Al-Degs et al. 2001). With increasing knowledge of the nature and genetic control of different diatom nanostructures will come the possibility of their manipulation for nanotechnological applications and their use in microelectronic devices, chemical and biological sensing and diagnosis, drug delivery, catalysis and nanometallurgy (see review by Bozarth et al. 2009). The possibility for compartmentalisation of different cellular processes offered by the nanostructure of the diatom frustule allows, for example, separation of biochemical synthesis from subsequent processing (Kroth 2007).

A range of molecular methods are now available for the identification and selection of required cyanobacterial strains for different biotechnological applications (Saker et al. 2009). DNA amplification processes and the generation of DNA profiles have been used to identify appropriate genotypes even under conditions which may display phenotypic variations such as those that are known to occur under culture conditions. The application of these techniques will provide high-level discrimination in the detection of, for example, appropriate target organisms for synthesis of high-value products. Recombinant DNA techniques have opened up the possibility of cyanobacterial strain improvement to suit specific commercial purposes. One interesting application that has been suggested (Boussiba et al. 2000) is the use of appropriately engineered cyanobacteria for mosquito control. The procedure would involve replacement of the currently used biological control method for mosquito larvae, the bacterium *Bacillus thuringensis* (Bti), by expressing the bacterial toxin gene in recombinant cyanobacteria as the delivery system. Such transgenic clones of *Anabaena* PCC 7120 with high toxicities and stability have indeed been produced although the problem of achieving effective transformation on a large scale remains as also does the danger, common to all such procedures, of releasing such recombinant strains to the environment. Loss of fitness of the transgenic strains, compared with the wild type, would be another likely impediment to future applications.

The main consideration for the various commercial applications based on the processing of microalgal biomass (for biofuel, bulk food or food-additive production) is the need for high sustainable yield in terma of volumetric or areal production combined with maximum cost-effectiveness. Special attention will inevitably focus on the faster-growing strains such as the high-temperature strain of *Chlorella* first described by Sorokin (1959) and reported to be capable of maintaining doubling times of *ca.* 3h, compared with the more usual 5 to 6h for this microalgal genus. Some diatom strains such as those isolated from shallow, estuarine locations, normally subject to both high temperatures and high irradiances, have also been shown to have high specific growth rates (Thompson et al. 1992). Cultures of *Chaetoceros salugineum* isolated from the Shinkawa-Kasugagawa estuary, Takamatsu, Japan were shown by Ichimi et al. (2012) to have very high growth potential under optimal conditions with daily specific growth rates of 8.4 d^{-1} (ln units) under continuous illumination.

One of the major constraints to rapid growth in culture is the efficient utilization of solar energy and the problems associated with scaling-up from the yields obtainable in the laboratory or at the pilot-plant level to those necessary to sustain large-scale production. Grobbelaar (2009) has calculated the upper limits of microalgal photosynthetic production as *ca.* 200g (dw) $m^{-2}d^{-1}$ at a photosynthetic efficiency of 8% (the equivalent of an annual biomass production of 700 $t.ha^{-1}$). The closest to these yield values attained with any large-scale open pond system is the 27g(dw).$m^{-2}d^{-1}$ for *Spirulina platensis*, the 25g(dw).$m^{-2}d^{-1}$ for *Chlorella* sp. (reported in Lee 2001) and the 20g(dw).$m^{-2}d^{-1}$ for *Dunaliella* (Borowitzka 1999), with photosynthetic efficiencies scarcely more than 1%. Considerably higher productivity values expressed as equivalent land area foot-print per day, have been recorded with various types of enclosed photobioreactor systems but, as Grobbelaar (2012) has pointed out, open raceway ponds represent the only practical large-scale intensive production system currently in wide use in various parts of the world. The same author has, however, predicted that current technology should allow average production rates close to 50g(dw) $m^{-2}.d^{-1}$ with a photosynthetic efficiency of *ca.* 2%, still well short of the 5% conversion efficiency predicted by Schenk et al. (2008). One of the key limitations of the open raceway system is mixing, a limitation that would require careful attention to pond design and extensive fluid dynamic modelling.

For any mass culturing system based on photosynthetic production, the demand for CO_2 is clearly crucial with a stoichiometric requirement of *ca.*1.7 $kgCO_2$ per kg dry biomass. This value will vary depending on the nature of the major product but for optimum photosynthetic yield CO_2 must be provided at partial pressures in excess of 1%. Control of CO_2 supply, and of other growth requirements can best be achieved in closed photobioreactors. Among the other advantages of such closed bioreactors is the possibility of providing an increased area for exposure to light (Pulz 2001) and ensuring the optimum (100μE.$m^{-2}s^{-}1$) exposure for adequate growth rates. Closed systems also avoid evaporation, allowing deployment in arid regions otherwise unsuitable for traditional agriculture or plant biomass production (Benemann & Oswald 1996). Competition from undesired microalgal species and losses from grazing are also more easily controlled as also are the essential mixing and harvesting requirements. Closed systems would also be most amenable to microalgal bimass production from mixotrophic or heterotrophic culture.

Considerations of the economics of microalgal biomass production systems, open or closed, inevitably raise comparisons with agricultural cropping systems. Microalgal systems, especially closed systems, have high equipment and maintenance costs and, as in modern agriculture generally, extensive additional energy requirements. Successful commercial application of microalgal culture systems will require the development of cheaper and more energy-efficient designs and the adoption of new technologies. Without them, production systems based on microlagal culture cannot compete with, for example, the use of fossil fuel or the processing of plant biomass into cheap fuel via fermentation. The former is, however, a finite resource with increasing costs of recovery while the latter competes with food production and a limited (and decreasing) land area suitable for efficient crop production. Microalgal culture methods have the advantage that, unlike plant crops (such as oil seeds or grains), the entire biomass is utilizable allowing greater yield per agricultural area. This, allied to their metabolic flexibility and their suitability for genetic manipulation to yield specific products, will in future form part of any large-scale adoption of microalgal culture systems.

The well-documented contribution of microalgae, over geological time scales, to the world's existing but ever dwindling easily extractable reserves of fossil fuels, points to the superficially attractive logic of using cultured microalgae as a direct source of renewable fuel. Any assessment of the economics of biofuel production from microalgae inevitably leads to comparisons of the cost of establishing a new industry with new and relatively untried technologies against that of well established extractive industries able to tap into available bulk resources. The case for sourcing even a small proportion of our biofuel requirement from microalgal culture is best made, as is the case for biomass generally, on the basis of continuity of supply. The main advantage of microalgal culture over crop-based biomass production relates to its suitability for marginal land, for various aquatic sites including coastal regions, waste waters and industrial effluents and to its tolerance of enhanced CO_2 sources such as industrial flue gases where the cost of production can be set against possible bioremediation effects. Another important feature of microalgal culture relates to the demonstrated ability of certain species to direct their metabolism toward the synthesis of specific products suitable as biofuel substrates.

The major challenges will come with attempts to scale up from the present conceptual low efficiency lab scale cultures to large scale outdoor efficient practical applications. Some of the ensuing problems are those intrinsic to any mass culture photosynthetic process such as the development of high O_2 tensions. Other problems relate to maintaining an appropriate balance between excesss photon supply (and loss of energy as heat or fluorescence) at the surface and light starvation in the body of the culture. Solutions for such problems may come from the use of mutant microalgal strains with smaller antennae sizes (Nakajima & Ueda 1997, 1999) and improved productivities in outdoor mass cultures.

Of all the biofuel applications based on microalgal photosynthetic productivity perhaps the most exciting, yet technically the most demanding, is that exploiting the ability of microalgae to carry out reactions leading to the production of hydrogen. Based on the considerable amount of research on the biochemistry of the processes involved, Benemann (2000) has carried out a prelinary practical analysis of how sustained microalgal hydrogen production might be achieved. He assumed a first stage of photosynthetic CO_2 fixation in large ponds of several hectares, fitted with raceway paddle wheels and other reqirements to maximise photosynthetic production. Harvested and concentrated microalgal biomass would then be supplied to a large anaerobic fermentor to induce hydrogenase and initiate H_2 production (and other useful biproducts such as acetate) perhaps by way of dark fermentation. Further cost savings might be achieved by linking such a system with an integrated wastewater treatment process where H_2 production would be a byproduct of waste water treatment.

The microalgae could be described as occupying an intermediate position between higher plants on the one hand and the microbiological world on the other. They share with the former a capacity for photosynthetic production and with the latter their fast, potentially exponential rates of growth and a degree of metabolic flexibility. Agricultural practices which, throughout man's history, have been directed towards improved yield of specific crops and plant products are, with few exceptions, yet to be applied to microalgae and their products. The microalgae could therefore be regarded as a largely untapped resource, one that, despite recent and continuing biotechnological advances, is yet to be fully exploited. A major driving force in this will be community demands for greater emphasis on renewable resources and an increasing awareness of the need to supplement the high energy demanding modern agriculture and industry with more environmentally acceptable processes.

Despite the considerable growth in recent years in the development of renewable energy sources, the world's large-scale power generation will continue to be dominated by sources based on the burning of fossil fuels. The potential contribution of microalgal technology to carbon capture (and other bioremediation processes) will thus become increasingly important. It is significant that a panel of economists reporting to the Copenhagen Consensus on Climate Change recommended that the best long-term strategy to address climate change would be to increase substantially investment in green research and technology. Continued expansion of research on the practical application of various aspects of microalgal technology will surely be central to this.

References

Abdulquader, G., Barsanti, L. & Tredici, M. R. (2000). Harvest of *Arthrospira platensis* from Lake Kossorom (Chad) and its household usage among the Kanembu. *J. Appl. Phycol.* 12:493-498.

Abed, R. M. M. & Golubic, S. (2009). Selection of oil-tolerant cyanobacteria on oil polluted sediments. *Algological Studies* 130:69-79.

Abed, R. M. M. & Köster, J. (2005). The direct role of aerobic heterotrophic bacteria associated with cyanobacteria in the degradation of oil compounds. *Int. Biodeter. & Biodeg.* 55:29-37.

Abeles, F. B. (1964). Cell-free hydrogenase from *Chlamydomonas. Plant Physiol.* 39:169-176.

Abrahamsson, K., Lorén A., Wulff, A. & Wängberg, S. Å. (2004). Air-sea exchange of halocarbons: the influence of diurnal and regional variations and distribution of pigments. *Deep-Sea Res. Part II: Tropical Studies in Oceanography* 51:2789-2805.

Abul- Hajj, Y. J. & Qian, X. (1986). Transformation of steroids by algae. *J. Nat. Prod.* 48:244-248.

Adams, M.W.W. (1990). The structure and mechanism of iron hydrogenases. *Biochim. Biophys Acta* 1020:115-145.

Adams, M. W. W., Mortenson, L. E. & Chen, J-S. (1981). Hydrogenase. *Biochim. Biophys. Acta*, 594:105-176.

Ahlgren, G., Gustafsson, I-B. & Boberg, M. (1992). Fatty acid content and chemical composition of freshwater microalgae. *J. Phycol.* 28:37-50.

Ahmed, H. & Häder, D-P. (2010). A fast algal bioassay for assessment of copper toxicity in water using *Euglena gracilis*. *J. Appl. Phycol.* 22:785-792.

Al-Degs, Y., Khraisheh, M. A. & Tutunji, M. F. (2001). Sorption of lead ions on diatomite and manganese oxides modifies diatomite. *Water Res.* 35:3724-3728.

Alderson, V. C. (1919). The Oil Shale Industry. *Quarterly of the Colorado School of Mines,* Golden, Colorado, April, 1918.

Alfinito, S., Iberite, M. & Fumanti, B. (1990). The algal microflora of the salt works of Tarquinia (Italy). *Hydrobiologia* 203:137-146.

Al-Hasam, R. H., Ail, A. M., Ka'wash, H. H. & Radwan, S. S. (1990). Effect of salinity on the fatty acid composition of the halophyte *Navicula* sp.: potential in mariculture. *J. Appl. Phycol.* 2:215-222.

Alldredge, A. L., Passow, U. & Logan, B. E. (1993). The abundance and significance of a class of large transparent organic particles in the ocean. *Deep-Sea Res.* I 40:1131-1140.

Allgaier, M., Vogt, M., Thyrhaug, R., Riebesell, U. & Grossart, H. P. (2008). Coupling of heterotrophic bacteria to phytoplankton bloom development at different pCO_2 levels: a mesocosm study. *Biogeosciences* Spec. Issue: (Pelagic ecosystem CO_2 enrichment studies) 5:317-359.

Almeida, C., Vilas, J. C., Martel, A., Suáres, S. & Reina, G. G. (2009). Microalgal strain selection for biodiesel production in a simple and low cost photobioreactor design. *Phycologia* 48. Abstract 7, 9th Int. Phycol. Cong. Tokyo.

Aluwihare, L. I. & Repeta, D. J. (1999). A comparison of the chemical characteristics of oceanic DOM and extracellular DOM produced by marine algae. *Mar. Ecol. Prog. Ser.* 186:105-117.

Amin, S. (2009). Review of biofuel oil and gas production processes from microalgae. *Energy Conversion & Management* 50:1834-1840.

Anderson, D. M. & Lobel, P. S. (1987). The continuing enigma of ciguatera. *Biol. Bull* 172:89-107.

Anderson, N. J. (2000). Diatoms, temperature and climate change. *Eur. J. Phycol.* 35:307-314.

Anderson, D. M., Kulis, D. M., Sullivan, J. J., Hall, S. & Lee, C. (1990). Dynamics and physiology of saxitoxin production by the dinoflagellates *Alexandrium* spp. *Mar. Biol.* 104:511-524.

Antia, N. J. & Cheng, J. Y. (1982). The keto-carotenoids of two marine coccoid members of the Eustigmatophyceae. *Br. Phycol. J.* 17:39-50.

Apple, J. K., Del Giorgio, P. A. & Kemp, W. M. (2006). Temperature regulation of bacterial production, respiration, and growth efficiency in a temperate salt-marsh estuary. *Aquat. Microb. Ecol.* 43:243-254.

Araz, M. K., Lee, C. H. & Lal, A. (2003). Ultrasonic separation in microfluidic capillaries. In: *IEEE Ultrasonic Symposium,* pp. 1066-1069.

Archer, S. D., Ragni, M., Webster, R., Airs, R. L & Geider, R. J. (2010). Dimethyl sulfoniopropionate and dimethyl sulphide production in response to photoinhibition in *Emiliania huxleyi. Limnol. Oceanigr.* 55:1579-1589.

Armbrust, E. V. (2009). The life of diatoms in the world's oceans. *Nature* 459:185-192.

Arnosti, C., Grossart, H. P., Mühling, M., Joint, I. & Passow, U. (2011). Dynamics of extracellular enzyme activities in seawater under changed atmospheric pCO_2 : a mesocosm investigation. *Aquat. Microb. Ecol.* 64:285-298.

Arrigo, K. R.., Perovich, D. K., Pickart, P. S., Brown, Z. W. and others. (2012). Massive phytoplankton blooms under Arctic sea ice. *Science* 336:1408.

Atsumi, S., Higashide, W. & Liao, J. C. (2009). Direct photosynthetic recycling of carbon dioxide to isobutyraldehyde. *Nat. Biotechnol.* 27:1177-1180.

Ayehunie, S., Belay,, A., Hu, Y., Baba, T. & Ruprecht, R. (1996). Inhibition of HIV-1 replication by an aqueous extract of *Spirulina platensis* (*Arthrospira platensis*). 7th IAAAA Cinference, Knysna, South Africa April 17.

Azizullah, A., Richter, P. & Häder, D-P. (2012). Sensitivity of various parameters in *Euglena gracilis* to short-term exposure to industrial wastewaters. *J. Appl. Phycol.* 24:187-200.

Badger, M. R., Andrews, T. J., Whitney, S. M., Ludwig, M., Yellowlees, D. C., Leggat, W. & Price, G. D. (1998). The diversity and coevolution of Rubisco, plastids, pyrenoids, and chloroplast-based CO_2-concentrating mechanisms in algae. *Can. J. Bot.* 76:1052-1071.

Bailey, D., Mazurak, A. P. & Rosowski, J. R. (1973). Aggregation of soil particles by algae. *J. Phycol.* 9:99-101.

Banerjee, A., Nyak, D., Chakrabotty, D. & Lahin, S. (2008). Uptake studies of environmentally hazardous ^{51}Cr in Mung beans. *Env. Pollution.* 151:423-427.

Banerjee, M. (2007). Comparative study on phosphatase activity of four rice field cyanobacterial strains and their ecological implications. *Nova Hedwigia* 85:407-416.

Banerjee, M. & John, J. (2005). Phosphatase activity of non-hair-forming cyanobacterium *Rivularia* and its role in phosphorus dynamics of deepwater rice fields. *Appl. Ecol. & Environ. Res.* 3:55-60.

Banerjee, A., Sharma, R., Chisti, Y. & Banerjee, U. C. (2002). *Botryococcus braunii*: a renewable source of hydrocarbons and other chemicals. *Crit. Rev. Biotechnol.* 22:245-279.

Bang, H. O., Dyerberg, J. & Sinclair, H. M. (1980). The composition of the Eskimo food in north western Greenland. *Am. J. Clin. Nutr.* 33:2657-2661.

Baos, R., García-Villada, L., Agrelo, M., López-Rodas, V., Hiraldo, F. & Costas E. (2002). Short-term adaptation of microalgae in highly stressful environments: an experimental model analysing the resistance of *Scenedesnus intermedius* (Chlorophyceae) to the heavy metals mixture from the Aznalcóllar mine spill. *Eur. J. Phycol.* 37:593-600.

Bar, E., Rise, M., Vishkautsan, M. & Shoshana, (M). R. (1995). Pigment and structural changes in *Chlorella zofingiensis* upon light and nitrogen stress. *J. Plant Physiol.* 146:527-534.

Barclay, W. R. & Lewin, R. A. (1986). Production of polysaccharides by microalgae on soil. *Plant and Soil* 88:159-169.

Barkay, T., Miller, S. M. & Summers, A. O. (2003). Bacterial mercury resistance from atoms to ecosystems. *FEMS Microbiology Reviews* 27:355-384.

Barkoh, A., Paret, J.M., Lyon, D. D., Begley, D. C., Smith, D. G. & Schlechte, J. W. (2008). Evaluation of barley straw and a commercial probiotic for controlling *Prymnesium parvum* in fish production ponds. *North American J. Aquaculture* 70:80-91.

Barkoh, A., Smith, D. G. & Schlechte, J.W. (2003). An effective minimum concentration of un-ionized ammonium nitrogen for controlling *Prymnesium parvum. North Azmerican J. Aquaculture* 65:220-225.

Barnett, T. P., Pierce, D. W., AchutaRao, K. M., Gleckler, P. J., Santer, B. D., Gregory, J. M. & Washington, W. M. (2005). Penetration of human-induced warming into the world's oceans. *Science* 309:284-287.

Barton, A. D., Dutkiewicz, S., Flier, G., Bragg, J. & Follows, M. J. (2010). Patterns of diversity in marine phytoplankton. *Science* 327:1509-1511.

Battarbee, R. W. & Charles,D. F. (1987). The use of diatom assemblages in lake sediments as a means of assessing the timing, trends and causes of lake acidification. *Prog. Phys. Geog.* 11:552-580.

Battarbee, R. W., Smol, J. P. & Meriläinen, J. (1986). Diatoms as indicators of pH: a historical review. In: Smol, J. P., Battarbee, R. W., Davies, R. B. & Meriläinen, J (eds.) *Diatoms in Lake Acidity.* W. Junk, The Hague, The Netherlands.

Beaugrand, G., Edwards, M. & Legendre, L. (2010). Marine biodiversity, ecosystem functioning and carbon cycles. *Proc. Natl. Acad. Sci. USA* 107:10120-10124.

Becher, P. G., Baymann, H. I. & Gademann, K. (2009). The cyanobacterial alkaloid nostocarboline: an inhibitor of acetylcholinesterase and trypsin. *J. Appl. Phycol.* 21:103-110.

Beeton, A. M. (1969). Changes in the environment and biota of the Great Lakes. In: *Eutrophication: causes, consequences, correctives. Proc. Nat. Acad. Sci. USA* pp. 150-187.

Belkin, I. M. (2009). Rapid warming of large marine ecosystems. *Prog. Oceanogr.* 81:207-213.

Ben-Amotz, A. & Avron, M. (1983). On the factors which determine the massive β-carotene accumulation in the halotolerant alga *Dunaliella bardawii. Plant Physiol.* 72:593-597.

Ben-Amotz, A., Tornabene, T. G. & Thomas, W. H. (1985). Chemical profile of selected species of microalgae with empbasis on lipids. *J. Phycol.* 21:72-81.

Benemann, J. R. (2000). Hydrogen production by microalgae. *J. Appl. Phycology* 12:291-300.

Benemann, J. R. & Oswald, W. J. (1996). Systems and economic analysis of microalgae ponds for conversion of carbon dioxide to biomass. *Pittsburgh Energy Technology Centre, Pittsburgh, PA., USA.* http://govdocs.aquake.org/cgi/reprint/2004/915/9150050.

Bennion, H. (1994). A diatom-phosphorus transfer function for shallow, eutrophic ponds in southeast England. *Hydrobiologia* 275/6:391-410.

Berday, N., Zaoui, D., Lamrini, A., & Bi, M. (2005). Impact of feeding activity of silver carp on plankton removal from a high-rate pond effluent. *Water Qual. Res. J. Can.* 40:191-201.

Berge, T., Daugbjerg, N., Andersen, B. B. & Hansen, P. J. (2010). Effect of lowered pH on marine phytoplankton growth rates. *Mar. Ecol. Prog. Ser.* 416:79-91.

Berger, W. H. (1982). Increase of carbon dioxide in the atmosphere during deglaciation: the coral reef hypothesis. *Naturwissenschaften* 69:87-88.

Bergman, B., Gallon, J. R., Rai, A. N. & Stal, L. J. (1997). N_2 fixation by non-heterocystous cyanobacteria. *FEMS Microbiol Rev.* 19:139-185.

Berry, J. P., Gibbs, P. D. L., Schmale, M. C. & Saker, M. L. (2009). Toxicity of cylindrospermopsis, and other apparent metabolites from *Cylindrospermopsis raciborskii* and *Aphanizomenon ovalisporum* to the zebrafish (*Danio rerio) embryo. Toxicon* 53:289-299.

Beuzenberg, V., Mountfort, D., Holland, P., Shi, F. & MacKenzie, L. (2011). Optimization of growth and production of toxins by three dinoflagellates in photobioreactor cultures. *J. Appl. Phycol.* DOI 10.1007/s10811-011-9726-8.

Bhattacharya, D. & Medlin, L. K. (1995). The phylogeny of the plastids, a review based on comparison of small subunit ribosomal RNA coding regions. *J. Phycology* 31:489-498.

Bhattacharya, P. & Pal, R. (2011). Response of cyanobacteria to arsenic toxicity. *J. Appl. Phycol.* 23:293-299.

Biopact. (2007). An in-depth look at biofuels from algae Fromhttp://biopact.com/2007/01/in depth-look-at biofuels-from-algae-html.

Birks, H. J. B., Line, J. M., Juggins, S., Stevenson, A. C. & ter Braak, C. J. F. (1990). Diatoms and pH reconstruction. *Phil. Trans. Roy. Soc. Lond.* B 327:263-278.

Blanc, G., Duncan, G and others. (2010). The *Chlorella variabilis* NC64A genome reveals adaptation to photosymbiosis, coevolution with viruses, and cryptic sex. *The Plant Cell* 22:2943-2955.

Blanck, H. & Wängberg, S-A. (1988). The validity of an ecotoxicological test system. Short-term and long-term effects of arsenate on marine periphytic communities in a laboratory system. *Ca. J. Fish. Aquat. Sci* 45:1807-1815.

Blevins, W. T., Schrader, K. K. & Saadoun, I. M. K. (1995). Comparative physiology of geosmin production by *Streptomyces halstedii* and *Anabaena* sp.*Water Sci. & Technol.* 31:127-133.

Boïchenko, V. A. & Hoffmann, P. (1994). Photosynthetic hydrogen production in Prokaryotes and Eukaryotes: Occurrence, mechanism and functions. *Photosynthetica* 30:527-552.

Bonnard, I., Rolland, M., Salmon, J-M., Debiton, E., Barthomeuf, C. & Banaigs, B. (2007). Total structure and inhibition of tumor cell proliferation of laxaphycins. *J. Med. Chem.* 50:1266-1279.

Booma, M., Selke, S. E. & Giacin, J. R. (1994). Degradable plastics. *J. Elastomers Plast.* 26:104-142.

Borde, X., Guieysse, B., Delgado, O., Muñoz, R., Hatti-Kaul, R., Nugier-Chauvin, C., Patin, H. & Mattiasson, B. (2003). Synergistic relationships in algal-bacterial microcosms for the treatment of aromatic pollutants. *Bioresour. Technol.* 86:293-300.

Borkenstein, C. G., Knoblechner, L., Frühwirth, H. & Schagerl, M. (2011). Cultivation of *Chlorella emersonii* with flue gas derived from a cement plant. *J. Appl. Phycol.* 23:131-135.

Borowitzka, L. J. (1991). Development of Western biotechnology algal beta-carotene plant. *Bioresource Technol.* 38:251-252.

Borowitzka, L. J. (1992). Commercial *Dunaliella* production: history of development. In: Villa, T. G & Abalde, J. (Eds.). *Profiles on Biotechnology.* Servicio de publicaciones, Universidad Santiago de Compostela, A Coruña, pp. 233-245.

Borowitzka, L. J., Borowitzka, M. A. & Moulton, T. P. (1984). The mass culture of *Dunaliella salina* for fine chemicals:from laboratory to pilot plant .*Hydrobiologia*116/117:115-121.

Borowitzka, M. A. (1995). Microalgae as sources of pharmaceuticals and other biologically active compounds. *J. Appl. Phycol.* 7:3-15.

Borowitzka, M. A. (1999). Commercial production of microalgae: ponds, tanks, tubes and fermenters. *J. Biotechnol.* 70:313-321.

Borowitzka, M. A., Borowitzka, L.J. & Kessly, D. (1990). Effect off salinity increase on carotenoid accumulation in the green alga *Dunaliella salina. J. Appl. Phycol.* 2:111-119.

Borowitzka, M. A., Huisman, J. M. & Osborn, A. (1991). Culture of the astaxanthin-producing green alga *Haematococcus pluvialis.* I. Effects of nutrients on growth and cell type. *J. Appl.. Phycology* 3:295-304.

Borowitzka, M. A. & Volcani, B. E. (1978). The polymorphic diatom *Phaeodactylum tricornutum.* Ultrastructure of its morphotypes. *J. Phycol.* 14:10-21.

Bosak, T., Liang, B., Sim, M. S. Petroff, A. P. (2009). Morphological record of oxygenic photosynthesis in conical stromatolites. *Proc. Natl. Acad. Sci.USA* 106:10939-10943.

Bosma, R., van Spronsen, W., Tramper, J. & Wijffels, R. (2003). Ultrasound, a new separation technique to harvest microlagae. *J. Appl. Phycol.* 15:143-153.

Boura-Halfon, S., Rise, M., Arad, S(M). & Sivan, A. (1997). Characterization of mutants of the red microalga *Porphyridium aerugineum* (Rhodophyceae) resistant to DCMU and atrazine. *Phycologia* 36:479-487.

Boussiba, S. (2000). Carotenogenesis in the green alga *Haematococcus pluvialis*: Cellular physiology and stress response. *Physiologia Plantarum,* 108:111-117.

Boussiba, S. & Vonshak, A. (1991). Astaxanthin accumulation in the green alga *Haematococcus pluvialis. Plant and Cell Physiol.* 32:1077-1082.

Boussiba, S. Wu, X-Q., Ben-Dov, E. Zarka, A. & Zaritsky, A. (2000). Nitrogen-fixing cyanobacteria as gene delivery system for expressing mosquitocidal toxins of *Bacillus thuringiensis* sp. *Israelensis. J. Appl. Phycol.* 12:461-467.

Bown, P. R., Burnett, J. A. & Gallagher, L. T. (1991). Critical events in the evolutionary history of calcareous nanoplankton. *Historical Biology* 5:279-290.

Bown, P. R. & Young, J. R. (1998). In: Bown, P. R. (Ed.) *Calcareous Nannofossil Biostratigraphy.* Kluwer Academic Publishers, Dordrecht, The Netherlands.pp.1 – 15.

Bozarth, A., Maier, U-G. & Zauner, S. (2009). Diatoms in biotechnology: modern tools and applications. *Appl. Microbiol. Biotechnol.* 82:195-201.

Bradbury, J. P. (1988). A climate-limnological model of diatom succession for paleolimnological interpretation of varved sediments at Elk Lake, Monnesota. *J. Paleolimnol.* 1:115-131.

Brányiková, I., Maršálková, B., Doucha, J., Brányik, T., Bišová, K., Zachleder, V. & Vitová, M. (2011). Microalgae – novel highly efficient starch producers.*Biotechnol. Bioeng.* 108:766-776.

Brasier, M. D. (1980). *Microfossils.* Allen and Unwin, London.

Bręk-Laitenen, G. & Ojala, A. (2011). Grazing of heterotrophic nanoflagellates on the eukaryotic picoautotroph *Choricystis* sp. *Aquat. Microb. Ecol.* 62:49-59.

Briggs,M. (2004).Widescale biodiesel production from algae *Energy Bulletin,* October 3,2004..

Brimblecombe, P & Shooter, D. (1986). Photo-oxidation of dimethylsulphide in aqueous solution. *Mar. Chem.* 19:343-353.

Brown, M. R, (2002). Nutritional Value and use of microalgae in aquaculture. In: Cruz-Suárez et al (eds). *Avances en Nutrición Acuicola VI.* Memorias del VI Simposium Internacional de Nutrición Acuícola. 3 al 6 de Septiembre del 2002. Cancún, Quintana Roo, México.

Brown, M. R., Jeffrey, S. W., Volkman, J. K. & Dunstan, G. A. (1997). Nutritional properties of microalgae for mariculture. *Aquaculture* 151:315-331.

Brown, M. & Robert, R. (2002). Preparation and assessment of microalgal concentrates as feeds for larval and juvenile Pacific oyster (*Crassostrea gigas*). *Aquaculture* 207:289-309.

Brussaard, C. P. D., Gast, G. J., Van Duyl, F. C. & Riegman, R. (1996). Impact of phytoplankton bloom magnitude on a pelagic microbial food web. *Mar. Ecol. Prog. Ser.* 144:211-221.

Burkholder, J. M., Glasgow, H. B. & Deamer-Melia, N. (2001). Overview and present status of the toxic *Pfiesteria* complex (Dinpohyceae). *Phycologia* 40:186-214.

Burnett, J. A., Young, J. R. & Bown, P.R. (2000). Calcareous nanoplankton and global climate change. In: Culver, S. J. & Rawson, P. F. (eds.) *Biotic Response to Global Change.* Cambridge University Press, Cambridge, U.K.

Cabrera, T. & Hur, S. B. (2001). The nutritional value of live foods on the larval growth and survival of Japanese flounder, *Paralichthys olivaceus. J. Appl. Aquaculture* 11:35-53.

Cai, W. J., Hu, X. P., Huang, W. J. Murrell, M. C. and others (2011). Acidification of subsurface coastal waters enhanced by eutrophication. *Nat. Beosci.* 4:766-770.

Caire, G., Storni de Cano, M., Zaccaro de Mule, M. C., Palma, R. M. & Colombo, K. (1997). Exopolysaccharides of *Nostoc muscorum* (Cyanobacteria) in the aggregation of soil particles. *J. Appl. Phycol.* 9:249-253.

Caldeira, K. & Wickett, M. E. (2003). Anthropogenic carbon and ocean pH. *Nature* 425:365.

Cane, R. F. (1977). Coorongite, balkashite and related substances-an annoted bibliography. *Trans. R. Soc. S. Aust.* 101:153-164.

Cannell, R. J. P., Kellam, S. J, Owsianka A. M. & Walker, J. M. (1988). Results of a large scale screen of microalgae for the production of protease inhibitors. *Planta Med.* 54:10-14.

Capone, D. G., Burns, J. A., Montoya, J. P., Subramanian, A. and others. (2005). Nitrogen fixation by *Trichodesmium* sp., an important source of new nitrogen to the tropical and subtropical North Atlantic Ocean. *Glob Biogeochem. Cycles* 19, GB2024, doi:10.1029/2004GB002331.

Capone, D. G., O'Neil, J. M., Zehr, J & Carpenter, E. J. (1990). Basis for diel variation in nitrogenase activity in the marine planktonic cyanobacterium *Trichodesmium thiebauttii. Appl. Environ. Microbiol.* 56:3532-3536.

Capone, D. G., Zehr, J. P., Paerl, H. W., Bergman, B. & Carpenter, E. J. (1997). *Trichodesmium,* a globally significant marine cyanobacterium. *Science* 276:1221-1229.

Carmichael, W. W., Drapeau, C. & Anderson, D. M. (2000). Harvesting of *Aphanizomenon flos-aquae* Ralfs ex Born and Flash. Var. Flos-aquae (cyanobacteria) from Klamath Lake for human dietary use. *J. Appl. Phycol.* 12:585-595.

Carmichael, W. W. & Falconer, I. R. (1993). Diseases related to freshwater blue-green algal toxins, and control measures. In: Falconer, I. R. (ed) *Algal toxins in seafood and drinking water.* Cambridge Academic Press, Cambridge, pp 187-209. *

Carvalho, A., Meireles, L., & Malcata, F. X. (2006). Microalgal reactors: a review of enclosed system designs and performances. *Biotechnol. Prog.* 22:1490-1506.

Casadevall, E., Metzger, P. & Puech, M.-P. (1984). Biosynthesis of triterpenoid hydrocarbons in the alga *Botryococcus braunii. Terahedron Lett.* 25:4123-4126.

Cermeño, P., Dutkiewicz, S., Harris, R. P., Follows, M., Schofield, O. & Falkowski, P. G. (2008). The role of nutricline depth in regulating the ocean carbon cycle. *Proc. Natl. Acad Sci.* 105:20344-20349.

Charlson, R. J., Lovelock, J. E., Andreae, M. O. & Warren, S. G. (1987). Oceanic phytoplankton, atmospheric sulphur, cloud albedo and climate. *Nature* 326:655-661.

Chang, C. H.., King, P. W., Ghirardi, M. L. & Kim, K. (2007). Atomic resolution modelling of the ferredoxin:[FeFe] hydrogenase complex from *Chlamydomonas reinhardtii. Biophys. J.* 93:3034-3045.

Charalampopoulou, A., Poulton, A. J., Tyrrell, T. & Lucas, M. I. (2011). Irradiance and pH affect coccolithophore community composition on a transect between the North Sea and the Arctic Ocean. *Mar. Ecol. Prog. Ser.* 431:25-43.

Chen, H-C. & Melis, A. (2004). Localisation and function of SulP a nuclear-encoded chloroplast sulfate permease in *Chlamydomonas reinhardtii. Planta* 220:198-210.

Chen, C. Y., Yeh, K. L., Alsyah, R., Lee, D. J. & Chang, J. S. (2011). Cultivation, photobioreactor design and harvesting of microalgae for biodiesel production: a critical review. *Biores. Technol.* 102:71-81.

Cherng, J. Y. & Shih, M. F. (2005). Preventing dyslipidemia by *Chlorella pyrenoidosa* in rats and hamsters after chronic high fat diet treatment. *Life Sciences* 76:3001-3013.

Chesapeake Bay Foundation. (2004). Manure impact on rivers, streams and the Chesapeake Bay. Cheapeake Bay Foundation, Annapolis, MD Available via http://www.cbf.org/site/DocServer/0723manure report_noembargo. pdf?docID=2143

Chevalier, P. & de la Noüe, J. (1985). Wastewater nutrient removal with microalgae immobilized in carrageenan. *Enzyme. Microb. Technol.* 7:621-624.

Chisholm, S. W., Olson, R. J., Zettler, E. R., Goericke, R., Waterbury, J. B. & Welschmeyer, N. A. (1988). A novel free-living prochlorophyte abundant in the oceanic euphotic zone. *Nature* 334:340-343.

Chisti, Y. (2007). Biodiesel from microalgae. *Biotech. Adv.* 25:294-306.

Cho, C-W., Pham, T. P. T., Kim, S. and others (2009). Toxicity assessment of common organic solvents using a biosensor based on algal photosynthetic activity measurement. *J. Appl. Phycol.* 21:683-689.

Choi, G-G., Kim, B-H, Ahn, C-Y. & Oh, H-M. (2011). Effect of nitrogen limitation on oleic acid biosynthesis in *Botryococcus braunii. J. Appl. Phycol.* 23:1031-1037.

Choi, S. H. & Kim, S. G. (1998). Lipopolysaccharide inhibition of rat hepatic microsomal epoxide hydrolase and glutathione-S-transferase gene expression of nuclear factor kB activation. *Biochem. Pharmacol.* 56:1427-1436.

Chorus, I. & Bartram, J. (Eds.). (1999). *Toxic cyanobacteria in water.* WHO, E&FN Spon, Great Britain.

Christaki, U., Giannakourou, A., van Wambeke, F. & Grégori, G. (2001). Nanoflagellate predation on auto- and heterotrophic picoplankton in the oligotrophic Mediterranean Sea. *J. Plankton Res.* 23:1297-1310.

Chung, I. K., Beardall, J., Mehta, S., Sahoo, D. & Stojkovic, S. (2011). Using marine macroalgae for carbon sequestration: a critical appraisal. *J. Appl. Phycol.* 23:877-886.

Churro, C., Alverca, E., Sam-Bento, F and others. (2009). Effects of bacillamide and newly synthesised defivatives on the growth of cyanobacteria and microalgal cultures. *J. Appl. Phycol.* 21:429-442.

Ciferri, O. (1983). *Spirulina,* the edible microorganism. *Microbial Rev.* 47:551-578.

Claquin, P., Probert, I., Lefebvre, S. & Veron, B. (2008). Effects of temperature on photosynthetic parameters and TEP production in 8 species of marine microalgae. *Aquat. Microb. Ecol.* 51:1-11.

Clark, R. L. & Wasson, R. J. (1986). Reservoir sediments. In: De Deckker, P. & Williams, W. D. (eds.) *Limnology in Australia* pp.497-507. CSIRO, Melbourne and Dr. W. Junk, Dordrecht.

Clément, A., Seguel, M., Arzul, G., Guzmán, L. Alarcón, C. (2001). A widespread outbreak of a haemolytic, ichthyotoxic *Gymnodinium* sp. in southern Chile. In: Halleggraeff, G. M., Blackburn, S.I., Bolch, C.J. & Lewis, R. J. (eds.) *Harmful Algal Blooms 2000.* Intergovernmental Oceanographic Commission, UNESCO, Paris, pp.66-69.

Codd, G. A., Morrison, L.F.& Metcalf, J. S. (2005). Cyanobacterial toxins: risk management for health protection. *Toxicol. Appl. Pharm.*203:264-277.

Codd, G. A., Morrison, L. F., Nath, M., Sano, T. & Kaya, K. (2009). Extraction of cyanostatins and their analysis, with microcystins and anabaenopeptin A, in a 21-year archive of cyanobacterial bloom samples. *Algological Studies* 130:53-68.

Cohen, L., Kim, K., King, P., Seibert, M. & Schulten, K. (2005). Finding gas diffusion pathways in proteins: application to O_2 and H_2 transport in Cpl [Fe-Fe]-hydrogenase and the role of packing defects. *Structure* 13:1-9.

Cohen, Y., Krumbein, W.E. & Shilo, M. (1977). Solar Lake (Sinai). 2. Distribution of photosyhthetic microorganisms and primary production. *Limnol. Oceanogr.* 22:609-620.

Colvin, V. L. (2003). The potential environmental impact of engineered nanomaterials. *Nature Biotechnology,* 21:1166-1170.

Combaz, A. (1980). Les kérogènes vus au microscope. In. Durand, B. (Ed) *Kerogen: Insoluble Organic Matter from Sedimentary Rocks.* Ėditions Technip, Paris

Connor, W. E. & Connor, S. L. (1986). Dietary cholesterol and fat and the prevention of coronary heart disease: risks and benefits of nutritional change. In; Hallgren, B., Levin, Ö., Rössner, S. & Vessby, B (Eds.) *Diet and Prevention of Coronary Heart Disease and Cancer.* Pp.113-147. Raven Press, New York.

Cooksey, K. E., Guckert, J. B., Williams, G. A. & Callis, P. R. (1987). Fluorometric determination of the neutral lipid content of microalgal cells using Nile Red. *J. Microbiol. Methods* 6:333-346.

Corliss, J. O. (1984). The kingdom Protista and its 45 phyla. *Bio Systems* 17:87 126.

Cornford, C. (1998). Source rocks and hydrocarbons of the North Sea. In: Glennie, K. W. (Ed.). *Petroleum Geology of the North Sea. Basic concepts and recent advances. 4th edition.* Blackwell Science Ltd. Oxford.

Costas, E., Carrillo, E., Ferrero, L. M., Agrelo, M., Garcí-Villada, L., Juste, J. & López-Rodas, V. (2001). Mutations of algae from sensitivity to resistance against environmental selective agents: the ecological genetics of *Dictyosphaerium chlorelloides* under lethal concentrations of 3-(3,4-dichlorophenyl)-1,1-dimethylurea (DCMU) herbicide *Phycologia,*40:391-398.

Cottrell, M. T. & Suttle, C. A. (1995). Dynamics of a lytic virus infecting the photosynthetic marine picoflagellate *Micromonas pusilla. Limnol. Oceanogr.* 40:730-739.

Craggs, R., Davies-Colley, R. J. Tanner, C. C. & Sukias, J. P. S. (2003). Avanced ponds systems: performance with high rate ponds of different depths and areas. *Water Sci. Technol.* 48:259-267.

Craggs, R., Sutherland, D. & Campbell, H. (2012). Hectare-scale demonstration of high rate algal ponds for enhancing wastewater treatment and biofuel production. *J. Appl. Phycol.* 24:329-337.

Crutzen, P. J. (1992). Ultraviolet on the increase. *Nature* 356: 104-105.

Cumming, B. F., Smol, J. P., Kingston, J. C. and others. (1992). How much acidification has occurred in Adirondack region lakes (New York, USA) since preindustrial times? *Can. J. Fish. Aquat. Sci.* 49:128-141.

Cunningham, L., McMinn, A., Riddle, M., Stark, J. & Snape, I. (2000). The ecology of, and human impacts on, near-shore benthic marine algal mats in the Windmill Islands, Antarctica. The Second International Conference on Applications of Micro- and Meioorganisms to Environmental Problems: Program and Abstracts, Avalon Institute of Applied Science, Canada, p. 36.

Cunningham, L., Stark, J. S., Snape, I., McMinn, A. & Riddle, M. J. (2003). Effects of metal and petroleum hydrocarbon contamination on benthic diatom communities near Casey Station, Antarctica: An experimental Approach. *J. Phycol.* 39:490-503.

Danovaro, R., Corinaldesi, C., Dell'Anno, A., Fuhrman, J. A., Middleburg, J. J., Noble, R. T. & Suttle, C. A. (2010). Marine viruses and global climate change. *FEMS Microbiol. Rev.*34:1-42.

Davis, L. S., Hoffmann, J. P. & Cook, P. W. (1990). Seasonal succession of algal periphyton from a wastewater treatment facility. *J. Phycol.* 26:611-617.

Davis, R. B. (1987). Paleolimnological diatom studies of acidification of lakes by acid rain: An application of Quaternary science. *Quat. Sci. Rev.* 6:147-163.

Debro, L. H. & Ward, H. B. (1979). Antibacterial activity of freshwater green algae. *Planta Med.* 36:375-378.

De Filippis, L. F. & Pallaghy, C. K. (1976). The effect of sub-lethal concentrations of mercury and zinc on *Chlorella.* III. Development and possible mechanisms of resistance to metals. *Zeitschrift für Pflanzenphysiologie* 79:323-335.

De la Hoz, H., Ben-Zvi, A., Burrell, R. & McCaffret, W. (2010). Optimization of heterotrophic microalgal cultures. *Special Abstracts/J. Biotechnol.* 150s(2010)S1-S576.

de la Jara, A., Mendosa, H., Martel, A., Molina, C, Nordström, C., de la Rosa, V. & Diaz, R. (2003). Flow cytometric determination of lipid content of marine dinoflagellate *Crypthecodinium cohnii. J. Appl.* Phycol. 15:433-438.

De la Noue, J., Laliberte, G. & Proulx, D. (1992). Algae and wastewater. *J. Appl. Phycol.* 4:247-254.

Delaunay, F., Marty, Y., Moal, J. & Samain, J-F. (1993). The effect of monospecific algal diets on growth and fatty acid composition of *Pecten maximus* (L) larvae. *J. Exp. Mar. Biol. Ecol.* 173:163-179.

Del Campo, J. A., Moreno, J., Rodriguez, H., Vargas, M. A., Rivas, J. & Guerrero, M. G. (2000). Carotenoid content of chlorophycean microalgae: factors determining lutein accumulation in *Muriellopsis* sp. (Chlorophyta). *J. Biotechnol.* 76:51-59.

Del Campo, J. A., Rodrigues, H., Moreno, J., Vargas, M. Á., Rivas, J. & Guerrero, M. G. (2004). Accumulation of astaxanthin and lutein in *Chlorella zofingensis* (Chlorophyta). *Appl. Microbiol. Biotechnol.* 64:848-854.

Demmig-Adams, B. & Adams, W. W. III. (2002). Antioxidants in photosynthesis and human nutrition. *Science,* 298:2149-2153.

Deng, M. D. & Coleman, J. R. (1999). Ethanol synthesis by genetic engineering in cyanobacteria. *Appl. Environ. Microbiol.* 65:523-528.

Denman, K. (2008). Climate change, ocean processes, and iron fertilization. *Mar. Ecol. Prog. Ser.* 364:219-225.

de Philippis, R., Ena, A. & Paperi, R. (2000). Assessment of the potential of *Nostoc* strains from the Pasteur Culture Collection for the production of polysaccharides of applied interest. *J. Appl. Phycol.* 12:401-407.

Derenne, S., Largeau, C. & Hatcher, P. (1992). Structure of *Chlorella fusca* algaenan: relationship with ultralaminae in lacustrine kerogens; species- and environment-dependent variations in the composition of fossil ultralaminae. *Organic Geochemistry* 18:417-422.

De Souza, M. P. & Yoch, D. C. (1995). Comparative physiology of dimethyl sulphide production by dimethyl sulfoniopropionate lyase in *Pseudomonas doudorofii* and *Alcaligenes* sp. strain. *Appl. Env. Microbiol.* 61:3986-3991.

Dimitrov, K. (2007). Green fuel technologies: a case study for industrial photosynthetic energy capture. Brisbane, Australia.http://www.nanostring.net/Algae/CaseStudy.pdf.

Di Toppi, L. S. & Gabbrielli, R. (1999). Response to cadmium in higher plants. *Enciron. Exp. Bot.* 41:105-130.

Dodds, W. K., Gudder, D. A. & Mollenhauer, D. (1995). The ecology of *Nostoc. J. Phycol.* 31:2-18.

Doney, S. C., Fabry, V. J., Feely, R. A.,& Kleypass, J. A. (2009). Ocean acidification: the other CO_2 problem. *Ann. Rev. Mar. Sci.* 1:169-192.

Dote, Y., Sawayama, S., Inoue, S., Minowa, T. & Yokoyama, S. (1994). Recovery of liquid fuel from hydrocarbon-rich microalgae by thermochemical liquefaction. *Fuel* 73:1855-1857.

D'Ovidio, F., De Monte, S., Alvain, S., Dandoneau, Y. & Lévy, M. (2010). Fluid dynamical niches of phytoplankton types. *Proc. Natl. Acad. Sci. USA* 107:18366-18370.

Doucha, J. & Livansky, K. (1995). Novel outdoor thin-layer high density microalgal culture system: Productivity and operation parameters. *Algol. Stud.* 76:129-147.

Doucha, J. & Lívanský, K. (2009). Outdoor open thin-layer microalgal photobioreactor: Potential productivity. *J. Appl. Phycol.* 21:111-117.

Doucha, J. & Livanský, K. (2012). Production of high-density *Chlorella* culture grown in fermenters. *J. Appl. Phycol.* 24:35-43.

Doucha, J., Livanský, K., Kotrbáček, V. & Zachleder, V. (2009). Production of *Chlorella* biomass enriched by selenium and its use in animal nutrition: a review. *Appl. Microbiol. Biotechnol.* 83:1001-1008.

Doucha, J., Straka, F. & Lívanský, K. (2005). Utilization of flue gas for cultivation of microalgae (*Chlorella* sp.) in an outdoor open thin-layer photobioreactor. *J. Appl. Phycol.* 17:403-412.

Douskova, I., Doucha, J., Lívanský, K., Machat, J., Novak, P., Umysova, D., Zachleder, V. & Vitova, M. (2009). Simultaneous flue gas bioremediation and reduction of microalgal biomass production costs. *Appl. Microbiol. Biotechnol.* 82:179-185.

Drevon, C. A. (1990). n-6 and n-3 fatty acids – how much and which balance? *Näringsforskning* 34:56-61.

Dundas, I., Johannessen, O. M., Berge, G. & Heimdal, B. (1989). Toxic algal blooms in Scandinavian waters. *Oceanogr.* 2:9-14.

Dwyer, J. H., Navab, M., Dwyer, K. M., Hassan, K and others. (2001). Oxygenated carotenoid lutein and progression of early atherosclerosis: The Los Angeles atherosclerosis study. *Circulation* 103:2922-2927.

Dyni, J. R. (2003). Geology and resources of some world oil-shale deposits. *Oil Shale* 20:193-252.

Eady, R. R. (1996). Structure-function relationships of alternative nitrogenases. *Chem. Rev.* 96:3013-3030.

Eady, R. R. (2003). Current status of structure and function relationships of vanadium nitrogenase. *Coordin. Chem. Rev.* 237:23-30.

Eckardt, N. A. (2010). The *Chlorella* genome: big surprises from a small package. *The Plant Cell* 22:2924.

Ehhalt, D. & Prather, M. (2001). Atmospheric chemistry and greenhouse gases, p. 239-287. *In* J. T. Houghton, Ding, Y., Griggs, D. J. and others (eds.). Climate change 2001: the scientific basis. Cambridge University, Scientific Press.

Ehling-Schulz, M., Bilger, W. & Scherer, S. (1997). UV-B-induced synthesis of photoprotective pigments and extracellular polysaccharides in the terrestrial cyanobacterium *Nostoc commune. J. Bacteriol.* 179:1940-1945.

El-Bestawy, E.A., El-Salam, A. Z. A., El-Rahman, A. & Mansy, H. (2007). Potential use of environmental cyanobacterial species in bioremediation of lindane-contaminated effluents. *Int. Biodeterioration & Biodegradation.* 59:180-192.

Elliott, J. A., Jones, I. D. & Thackeray, S. J. (2006). Testing the sensitivity of phytoplankton communities to changes in water temperature and nutrient load in a temperate lake. *Hydrobiologia* 559:401-411.

Ergolu, E., Okada, S. & Melis, A. (2011). Hydrocarbon productivities in different *Botryococcus* strains: comparative methods in product quantification. *J. Appl. Phycol.* 23:763-775.

Eriksen, N. T. (2008). The technology of microalgal culturing. *Biotechnol. Lett.* 30:1525-1536.

Evans, C., Malin, G., Wilson, W.H. & Liss. P. S. (2006). Infectious titres of *Emiliania huxleyi* virus EhV-86 are reduced by exposure to millimolar dimethyl sulphide (DMS) and acrylic acid (AA). *Limnol. Oceanogr.* 51:2468-2471.

Fabregas, J., Herrero, C., Parafita, M., Paz, J. M., Cabezas, B. & Abalde, J. (1988). Deacrease in plasma cholesterol, triglycerides and CPK levels in rats fed on the marine microalga *Dunaliella tertiolecta. Planta Med.* 54:109-111.

Falconer, I. R. (2005). *Cyanobacterial Toxins of Drinking Water Supplies : Cylindrospermopsins and Microcystins*. CRC Press, Boca Raton, USA.

Falconer, I. R. & Buckley, T. H. (1989). Tumour promotion by *Microcystis* sp., a blue-green alga occurring in water supplies. *Medical Journal of Australia* 150:351.

Falkowski, P. G., Barber, R. T. & Smetacek, V. V. (1998). Biogeochemical controls and feedbacks on ocean primary production. *Science* 281:200-207.

Falkowski, P. G.,et al. (2005). The rise of oxygen over the past 205 million years and the evolution of large placental mammals. *Science* 309:2202-2204.

Falkowski, P. G. & Oliver, M. J. (2007). Mix and match: how climate selects phytoplankton. *Nature Reviews/Microbiology* 5:813-819.

Farhat, N., Rabhi, M., Falleh, H., Jouini, J., Abdelly, C. & Smaoui, A. (2011). Optimization of salt concentrations for a higher carotenoid production in *Dunaliella salina* (Chlorophyceae). *J. Phycol.* 47:1072-1077.

Farnelid, H., Tarangkoon, W., Hansen, G., Hansen, P. J. & Riemann, L. (2010). Putative N_2-fixing heterotrophic bacteria associated with dinoflagellate-Cyanobacteria consortia in the low-nitrogen Indian Ocean. *Aqauat. Microb. Ecology* 61:105-117.

Fedorov, A. S., Kosourov, S., Ghirardi, M. L. & Seibert, M. (2005). Continuous hydrogen photoproduction by *Chlamydomonas reinhardtii. Appl. Biochem. Biotechnol.* 121-124:403-412.

Feely, R. A., Sabine, C. L., Hernandez-Ayon, J. M., Ianson, D. & Hales, B. (2008). Evidence for upwelling of corrosive 'acidified' water onto the continental shelf. *Science* 320:1490-1492.

Felföldi, T., Duleba, M., Somogyi, B and others. (2011). Diversity and seasonal dynamics of the photoautotrophic picoplankton in Lake Balaton (Hungary). *Aquat. Microb. Ecol.* 63:273-287.

Feng, Y., Hare, C. E., Leblanc, K., Rose, J. M. and others (2009). Effects of increased pCO_2 and temperature on the North Atlantic Spring Bloom. I. The phytoplankton community and biogeochemical response. *Mar. Ecol. Prog. Ser.* 388:13-25.

Feofilova, E. P., Sergeeva, Ya.E. & Ivashechkin, A. A. (2010). Biodiesel-Fuel: Content, production, producers, conemporary biotechnology (Review). *Applied Biochemistry and Microbiology* 46:369-378.

Field, J. D. (1985). Organic geochemistry in exploration of the North Sea. In: Thomas, B. M., Doré, A. G., Eggen, S. S., Home, P. C. & Larsen, R. M. (eds.) *Petroleum Geochemistry and Exploration of the Norwegian Shelf.* Graham & Trotman, London, pp39-57.

Flores, H. S., Wikfors, G. H. & Dam, H. G. (2012). Reactive oxygen species are linked to the toxicity of the dinoflagellate *Alexandrium* sp. to protists. *Aquat. Microbial. Ecol.* 66:199-209.

Florin, G. A., Sanda, A., Cosmina, B., Marian, T., Pompei, B. & Cornel, C. (2011). Inhibition of chemically induced mammary and non-mammary carcinogenesis by astaxanthin in Wistar rats. *Planta Med.* 77

Florin, L., Tsokogolu, A. & Happe, T. (2001). A novel type of iron hydrogenase in the green alga *Scenedesmus obliquus* is linked to the photosynthetic electron transport chain. *J. Biol. Chem.* 276:6125-6132.

Forestier, M., King, P., Zhang, L., Posewitz, M., Schwarzer, S., Happe, T., Ghirardi, M. L. & Seibert, M, (2003). Expression of two [Fe]-hydrogenases in *Chlamydomonas reinhardtii* under anaerobic conditions. *Eur. J. Biochem.* 270:2750-2758.

Frankignoulle, M., Canon, C. & Gattuso, J. P. (1994). Marine calcification as a source of carbon dioxide: Positive feedback of increasing atmospheric CO_2. *Limnol. Oceanogr.* 39:458-462.

Franklin, D. J., Steinke, M., Young, J., Probert, I. & Malin, G. (2010). Dimethylsulphoniopropionate (DMSP), DMSP-lyase activity (DLA) and dimethylsulphide (DMS) in 10 species of coccolithophores. *Mar. Ecol. Prog. Ser.* 410:13-23.

Franklin, N. M., Stauber, J. L., Aote, S. C. & Lim, R. P. (2002). Effect of initial cell density on the bioavailability and toxicity of copper in microalgal bioassays. *Environ. Toxicol. Chem.* 21:742-751.

Frey, M. (2001). Radical mechanisms of enzymic catalysis. *Ann. Rev. Biochem.* 70:121-124.

Frey, M. (2003). Hydrogenase, hydrogen-activating enzymes. *Chembiochem.* 3:153-160.

Fritz, S. C., Juggins, S. & Battarbee, R. W. (1993). Diatom assemblages and ionic characteristics of lakes of the northern Great Plains, NA: A tool for reconstructing past salinity and climate fluctuations. *Can. J. Fish. Aquat. Sci.* 50:1844-1856.

Fu, F-X., Mulholland, M. R., Garcia, N. S. Beck, A and others. (2008). Interactions between changing pCO_2, N_2 fixation and Fe limitation in the marine unicellular cyanobacterium *Crocosphaera. Limnol & Oceanogr.* 53:2472-2484.

Fuhrman, J. A. (1999). Marine viruses and their biogeochemical and ecological effects. *Nature* 399:541-548.

Fujiwara, Y., Hirakawa, K. & Sinpo, K. (1990). Effect of long-term administration of *Chlorella* tablets on hyperlipemia. *J. Japanese Soc. Nitrition Food Science* 43:167-173

Fukao, T., Kimoto, K. & Kotani, Y. (2012). Effect of temperature on cell growth and production of transparent exopolymer particles by the diatom *Coscinodiscus granii* isolated from marine mucilage. *J. Appl. Phycol.* 24:181-186.

Fukao, T., Kimoto, K., Yamatogi, T., Yamamoto, K., Yoshida, Y. & Kotani, Y. (2009). Marine mucilage in Ariake Sound, Japan, is composed of transparent exoploymer particles produced by the diatom *Coscinodiscus granii. Fish Sci.* 75:1007-1014.

Fukuda, H., Kondo, A. & Noda, H. (2001). Biodiesel fuel production by transesterification of oils. *J. Biosci. Bioeng.* 92:405-416.

Fussel, H. M. (2009). An updated assessment of the risks from climate change based on research published since the IPCC Fourth Assessment Report on Climate Change 97:469-482.

Gaffron, H. & Rubin, J. (1942). Fermentative and photochemical production of hydrogen in algae. *J. Gen. Physiol.* 26:219-240.

Gallois, R. W. (1976). Coccolith blooms in the Kimmeridge Clay and origin of North Sea Oil. *Nature* 259:473-475.

Galván, F. & de la Rosa, F.F. (1997). Immobilized photosynthetic systems: Utilization of solar energy for production of chemicals and fuels. In: Pessarakei, M (Ed.) *Handbook of Photosynthesis.*:739-749. Marcel Dekker Inc. New York.

Gálves-Valdivieso, G., Cardeñosa, R., Vera, J. M., Pineda, M. & Aguilar, M. (2011). γ-Tocopherol methyltransferase from the green alga *Chlamydomonas reinhardtii*: functional characterisation and expression analysis. *Physiologia Plantarum* 143:316-328.

Gao, K., Wu, Y., Li, G., Wu, H., Villafañe, V. E. & Hebling, E. W. (2007). Solar UV radiation drives CO_2 fixation in marine phytoplankton: a double-edged sword. *Pl. Physiol.* 144:54-59.

Garcia, J., Mujeriego, R. & Hernández-Mariné, M. (2000). High rate algal pond operation strategies for urban wastewater nitrogen removal. *J. Appl. Phycol.* 12:331-339.

Garcia, S. M. & Rosenberg, A. A. (2010). Food security and marine capture fisheries: characteristics, trends, drivers and future perspectives. *Phil.Trans. Roy. Soc. Biol. Sci.* 365:2869-2880.

Garcia-González, M., Moreno, J., Manzano, J. C., Florencio, F. J. & Guerrero, M. G. (2005). Production of *Dunaliella salina* biomass rich in 9-*cis*-β-carotene and lutein in a closed tubular photobioreactor. *J. Biotechnol.* 115:81-90.0

Garcí-Malea, M. C., Acién, F.G., Del Río, E. and others (2009). Production of astaxanthin by *Haematococcus pluvialis*:taking the one-step system outdoors. *Biotechnol & Bioeng.* 102:651-658.

Garcia-Plazaola, J. I. & Becerril, J. M. (1999). A rapid high-performance liquid chromatography method to measure lipophilic antioxidants in stressed plants: simultaneous determination of carotenoids and tocopherols. *Phytochemical Analysis,* 10:307-313.

Gardner, R., Peters, P., Peyton, B. & Cooksey, K. E. (2010). Medium pH and nitrate concentration effects on accumulation of triacylglycerol in two members of the chlorophyta. *J. Appl. Phycol.* 22: DOI 10.1007/s10811-010-9633-4

Garson, M. J. (2001). Ecological perspectives on marine natural product biosynthesis. In: McClintock, J. B. & Baker, B. J. (Eds.). *Marine Chemical Ecology.* CRC Press, Boca Raton.

Geier, S. C., Huyer, S., Praebst, K., Husmann, M., Walter, C. & Buchholz, R. (2012). Outdoor cultivation of *Chlamydomonas reinhardtii* for photobiological hydrogen production. *J. Appl. Phycol.* 24:319-327.

Gell, P. A. (1993). Using salt lake diatoms for palaeosalinity reconstructions – Australia's contribution to the CASPIA Project. *Quat. Aust.* 11:33-34.

Gelpi, E., Oró, J., Schneider, H. J. & Bennett, E. O. (1968). Olefins of high molecular weight in two microscopic algae. *Science* (Wash. D.C.) 161:700-702.

Gernant, R. E., Gibson, T. G. & Whitmore, F. C. (1971). Environmental history of Maryland Miocene. *Maryland Geological Survey Guidebook No.3,* 58pp.

Gershwin, M. E. & Belay, A. (Eds.) (2008). *Spirulina in Human Nutrition and Health.* CRC Press, Boca Raton.

Ghasemi, Y., Rasoul-Amini, S. & Fotooh-Abadi, E. (2011). The biotransformation, biodegradation and bioremediation of organic compounds by microalgae. *J. Phycol.* 47:969-980.

Ghirardi, M. L. (2006). Hydrogen production by photosynthetic green algae. *Indian J. Biochem. & Biophys.* 43:201-210.

Ghirardi, M. L., Flynn, T., Forestier, M., Iyer, A., Melis, A., Danielson, P. & Siebert, M. (1998). Generation of *C. reinhardtii* mutants that photoproduce H_2 from H_2O in the presence of O_2. In: Garab, G. (Ed). *Proc. XIth International Congress on Photosynthesis, Budapest, Hungary.* Kluwer Academic Publishers, Dordrecht. pp. 1959-1962.

Ghirardi, M. L., Kosourov, S. & Tsygankov, A. (2000a). Two-phase photobiological algal H_2-production system. *Proc. 2000 DOE Hydrogen Program Review.*

Ghirardi, M. L., Zhang, L., Lee, J. W., Flynn, T., Siebert, M., Greenbaum, E. & Melis, A. (2000b). Microalgae: a green source of renewable H_2. *Trends Biotechnol.* 18:506-511.

Giacobbe, M. G., Penna, A., Ceredi, A., Milandri, A., Poletti, R. & Yang, X. (2000). Toxicity and ribosomal DNA of the dinoflagellate *Dinophysis sacculus* (Dinophyta). *Phycologia* 39:177-182.

Gibson, M. T., Welch, I. M.., Barrett, P. R. F. & Ridge, I. (1990). Barley straw as an inhibitor of algal growth II: Laboratory studies. *J. Appl. Phycology* 2:241-248.

Giglio, S., Saint, C. P. & Monis, P. T. (2011). Expression of the geosmin synthase gene in the cyanobacterium *Anabaena circinalis* AWQC318. *J. Phycol.* 47:1338-1343.

Gilbert, N. S. (1991). Primary production by benthic microalgae in nearshore marine sediments of Signy Island, Antarctica. *Polar. Biol.* 11:339-346.

Gilmore, S., Weston, P. & Thomson, J. (1993). A simple, rapid, inexpensive and widely applicable technique for purifying plant DNA. *Aust, Syst. Bot.* 6:139-142.

Gnanadesikan, A & Anderson, W. G. (2009). Ocean water clarity and the ocean general circulation in a coupled climate model. *J. Phys. Oceanogr.* 39:314-332.

Golakoti, T., Ogino, J., Heltzel, C. E. and others. (1995). Structural determination, conformational analysis, chemical stability studies, and antitumour evaluation of the cryptophycins. Isolation of 18 new analogs from *Nostoc* sp. strain GSV 224. *J. Amer. Chem. Soc.* 117:12030-12049.

Golden, J. W. & Yoon, H. S. (2003). Heterocyst development in *Anabaena. Curr. Opin. Microbiol.* 6:557-563.

Gondwe, M., Krol, M., Gieskes, W., Klaassen, W. & de Baar, H. (2003). The contribution of ocean-leaving DMS to the global atmospheric burdens of DMS, MSA, SO_2 and NSS $So_4^=$. *Global Biogeochem. Cyc.* 17:1056.

Gonzalez de Rivera, C., Miranda-Zamora, R., Diaz-Zagoya, J. C. & Juarez- Oropeza, M. A. (1993). Preventative effect of *Spirulina maxima* on the fatty liver induced by a fructose rich diet in rat, a preliminary report. *Life Sci.* 53:57.

González, C., Marciniak, J., Villaverde, S., García-Encina, P. A. & Muñoz, R. (2008). Microalgal-based processes for the biodegradation of pretreated piggery wastwaters. *Appl. Microbial. Biotechnol.* 80:891-898.

González, J. M. & Suttle, C. A. (1993). Grazing by marine nanoflagellates on viruses and virus-sized particles: ingestion and digestion. *Mar. Ecol. Prog. Ser.* 94:1-10.

Gordon, J. M. & Polle, J. E. (2007). Ultrahigh bioproductivity from algae. *Appl. Microbiol. Biotechnol.* 76:969-975.

Görs, M., Schumann, R., Hepperle, D. & Karsten, U. (2010). Quality analysis of commercial *Chlorella* products used as dietary supplements in human nitrition. *J. Appl. Phycol.* 22:265-276.

Gosselin, S., Fortier, L. & Gagné, J. A. (1989). Vulnerability of marine fish larvae to the toxic dinoflagellate *Protogonyaulax tamarensis*. *Mar. Ecol. Prog. Ser.* 57:1-10.

Goth, K., de Leeuw, J. W., Püttmann, W. & Tegelaar, E. W. (1988). Origin of Messel oil shale kerogen. *Nature* 336:759-761.

Gouveia, L., Gomes, E. & Empis, J. (1997). Use of *Chlorella vulgaris* in rainbow trout, *Oncorhynchus mykiss*, diets to enhance muscle pigmentation. *J. Appl. Aquaculture* 7:61-70.

Granéli, E. & Flynn, K. (2006). Chemical and physical factors influencing toxin content. In: Granéli, E., & Turned, J. T (Eds.). *Ecology of Harmful Algae, Ecological Studies* 189 pp.229-241. Springer-Verlag, Berlin, Heidelberg.

Granéli, E. & Turner, J. T. (2006). An introduction to harmful algae. In: Granéli, Turner, J. T. (Eds.) *Ecolofy of Harmful Algae,Ecological Stydies* 189:3-7.

Granum, E., Kirkvold, S. & Myklestad, S. M. (2002). Cellular and extracellular production of carbohydrates and amino acids by the marine diatom *Skeletonema costatum:* diel variations and effects of N depletion. *Mar. Ecol. Prog. Ser.* 242:83-94.

Gray, K. A., Zhao, L. & Emptage, M. (2006). Bioethanol. *Current Opinion in Chemical Biology* 10:141-146.

Greenbaum, E. (1988). Energetic efficiency of hydrogen photoevolution by algal water splitting. *Biophys. J.* 54:365-368.

Griffiths, D. A. & Griffiths, D. J. (1969). The fine structure of autotrophic and heterotrophic cells of *Chlorella vujlgaris* (Emerson strain). *Plant Cell Physiol.* 10:11-19.

Griffiths, D. J. (1965). The accumulation of carbohydrate in *Chlorella vulgaris* under heterotrophic conditions. *Ann. Bot.* N.S. 29:347-357.

Griffiths, D. J. & Saker, M. J. (2003). The Palm Island Mystery Disease 20 years on: a review of research on the cyanotoxin cylindrospermopsin. *Environ. Toxicol.* 18:78-93.

Griffiths, M. J. & Harrison, S. T. L. (2009). Lipid productivity as a key characteristic for choosing algal species for biodiesel production. *J. Appl. Phycol.* 21:493-507.

Griffiths, M. J., van Hille, R. P. & Harrison, S. T. L. (2011). Lipid productivity, settling potential and fatty acid profile of 11 microalgal species grown under nitrogen replete and limited conditions. *J. Appl. Phycol.*DOI 10.1007/s10811-011-9723-y

Grobbelaar, J. U. (2009). From laboratory to commercial production: a case study of a *Spirulina (Arthrospira)* facility at Musina, South Africa. *J. Appl.Phycol.* 21:523-527.

Grobbelaar, J. U. (2012). Microalgae mass culture: the constraints of scaling-up. *J. Appl. Phycol.* 24:315-318.

Grossart, H. P., Allgaier, M., Passow, U. & Riebesell, U. (2006). Testing the effect of CO_2 concentration on the dynamics of marine heterotrophic bacterioplankton. *Limnol. Oceanogr.* 51:1-11.

Grünewald, K., Hirschberg, J. & Hagen, C. (2001). Ketocarotenoid biosynthesis outside of plastids in the unicellular green alga *Haematococcus pluvialis*. *J. Biol. Chem.* 276:6023-6029.

Gudin, C. & Thepenier, C. (1986). Bioconversion of solar energy into organic chemicals by microalgae. *Adv. Biotechnol. Proc.* 6:73-110.

Guerin, M., Huntley, M. E. & &Olaizola, M. (2003). *Haematococcus* astaxanthin: applications for human health and nutrition. *Trends Biotechnol.* 21:210-216.

Guo, M., Harrison, P. J. & Taylor, F. J. R. (1996). Fish kills related to *Prymnesium parvum* N .Carter (Haptophyta) in the People's Republic of China. *J. Appl. Phycol.* 8:111-117.

Gupta, A. B. & Gupta, K. K. (1972). Effect of *Phormidium* extract on growth and yield of *Vigna catjang* (Cowpea) T.5269. *Hydrobiologia,* 41:127-132.

Gupta, A. & Sainis, J. K. (2010). Isolation of C-phycocyanin from *Synechococcus* sp., (*Anacystis nidulans* BD1). *J. Appl. Phycol.* 22:231-233.

Gustavson, K. & Wängberg, S-A. (1995). Tolerance induction and succession in microalgae communities exposed to copper and atrazine. *Aquatic Toxicology* 32:283-302.

Gutiérrez-Ruíz, M. E., Villalobos-Peñaloza, M. & Miranda, J. A. (1990). Chromium pollution in groundwaters in north of Mexico City. *Rev. Int. Contam. Amb.* 6:5-18.

Guzmán, H. M., Valido, A. de la J., Duarte, L. C. & Presmanes, K. F. (2010). Analysis of interspecific variation in relative fatty acid composition: use of flow cytometry to estimate unsaturation index and relative polyunsaturated fatty acid content in microalgae. *J. Appl. Phycol.* 18:189-199.

Guzzon A. & Albertano, P. (2009). Annual photosynthetic response of phototrophic biofilms from an Italian wastewater treatment plant. *Algalogical Studies* 131:87-102.

Gwizdala, M., Wilson, A. & Kirilovsky, D. (2011). *In Vitrio* reconstitution of the cyanobacterial photoprotective mechanism mediated by the orange carotenoid protein in *Synechocystis* PCC 6803. *The Plant Cell* 23:2631-2643.

Haase, S. M., Huchzermeyer, B. & Rath, T. (2012). PHB accumulation in *Nostoc muscorum* under different carbon stress situations. *J. Appl. Phycol.*24:157-162.

Hagmann, L. & Jüttner, F. (1996). Fischerellin A: A novel photosystem II-inhibiting allelochemical of the cyanobacterium *Fischerella muscicola* with antifungal and herbicidal activity. *Tetrahedron Lett.* 37:6539-6542.

Halac, S. R., Villafañe, V. E. & Helbling, E. W. (2010). Temperature benefits the photosynthetic performance of the diatoms *Chaetoceros gracilis* and *Thalassiosira weissflogii* when exposed to UVR. *J. Photochemistry and Photobiology B:Biology* 101:196-205.

Hallam, A. & Perch-Nielsen, K. (1990). The biotic record of events in the marine realm at the end of the Cretaceous: calcareous, siliceous and organic-walled microfossils and macroinvertebrates. *Tectonophysics* 171:347-357.

Hallegraeff, G. M. (2003). Harmful algal blooms: a global overview. In: Hallegraeff, G. M., Anderson, D. M., & Cembella, A. D. (Eds.) *Manual on harmful marine microalgae*, 11, IOC-UNESCO, Paris, pp25-49.

Hallenbeck, P. C., Kochian, L. V. & Benemann, J. R. (1981). Hydrogen evolution catalyzed by hydrogenase in cultures of cyanobacteria. *Z. Naturforsch.* 36C:87-92.

Hamada, H., Kondo, Y.,Ishihara, K., Nakajima, N.et al. (2003). Stereoselective biotransformation of Limonene and Limonene oxide by cyanobacterium, *Synechococcus* sp. PCC 7942. *J. Biosci. Bioeng.* 96:581-584.

Hanagata, N., Takeuchi, T., Fukuju, Y., Barnes, D. J. & Karube, I. (1992). Tolerance of microalgae to high CO_2 and high temperature. *Phytochem.* 31:3345-3348.

Happe, T. & Naber, J. D. (1993). Isolation, charcterisation and N-terminal amino acid sequence of hydrogenase from the green alga *Chlamydomonas reinhardtii. Eur. J. Biochem.* 214:475-481.

Harada, K-I., Ogawa, K., Matsuura, K., Marata, H. and others. (1990). Structural determination of geometrical isomers of microcystin-LR and –RR from cyanobacteria by two-dimensional NMR spectroscopic technique. *Chem. Res. In Toxicology* 3:473-481.

Harter, T and nine others. (2012). *J. Appl. Phycol.* DOI 10.1007/s10811-012-9870-9.

Harun, R., Danqual, M. K. & Forde, G. M. (2010). Microalgal biomass as a fermentation feedstock for bioethanol production. *J. Chem. Technol. & Biotechnol.* 85:199-203.

Harwood, J. L.& Guschina, I. A. (2009). The versatility of algae and their lipid metabolism. *Biochimie* 91:679-684.

Hawkins, P. R., Runnegar, M. T. C., Jackson, A. R. B. & Falconer, I. R. (1985). Severe hepatotoxicity caused by the tropical cyanobacterium *Cylindrospermopsis raciborskii* (Woloszynska) Seenaya and Subba Raju isolated from a domestic water supply reservoir. *Appl. Environ. Microbiol.* 50:1292-1295.

Hazewinkel, R. R. O., Wolfe, A. P., Pla, S., Curtis, C. & Hadley, K. (2008). Have atmospheric emissions from the Athabasca Oil Sands impacted lakes in northeastern Alberta, Canada? *Can. J. Fish. Aquat. Sci.* 65:1554-1567.

Hedges, J. I. (2002). Why dissolved organic matter? In: Hansell, D. A. & Carlson, C. A. (eds.) *Biogeochemistry of Marine Dissolved Organic Matter.* Academic Press, St. Louis, MO, pp1-33.

Henley, P., Hironaka, J. L., Guillou, L., Buchheim, M. A., Buchheim, J. A., Fawley, M. W. & Fawley, K. P. (2004). Phylogenetic analysis of the '*Nannochloris*-like' algae and diagnoses of *Picochlorum oklahomensis gen. et sp. nov.* (Trebouxiophyceae, Chlorophyta). *Phycologia* 43:641-652.

Hewson, I., O'Neil, J. M., Fuhrman, J. A. & Dennison, W. C. (2001). Virus-like particle distribution and abundance in sediments and overlying waters along eutrophic gradients in two subtropical estuaries. *Limnol. Oceanogr.* 46:1734-1746.

Heyer, H., Stal, L. J. & Krumbein, W. E. (1989). Simultaneous heterolactic and acetate fermentation in the marine cyanobacterium *Oscillatoria limosa* incubated anaerobically in the dark. *Arch. Microbiol.* 151:558-564.

Hill, J., Nelson, E., Tilman, D., Polasky, S. & Tiffany, D. (2006). Environmental, economic, and energetic costs and benefits of biodiesel and ethanol biofuels. *Proc. Natl. Acad. Sci. USA* 103:11206-11210.

Hirooka, T., Nagase, H., Hirata, K. & Miyamoto, K. (2006). Degradation of 2,4-dinitrophenol by a mixed culture of photoautotrophic microorganisms. *Biochem. Eng. J.* 29:157-162.

Hirschberg, J. (1999). Production of high-value compounds: carotenoids and vitamin E. *Current Opinion in Biotechnology,* 10:186-191.

Ho, A.Y.T., Hsieh, D. P. H. & Qian, P-Y. (2006). Variations in paralytic shellfish toxin and homolog production in two strains of *Alexandrium tamarense* after antibiotic treatments. *Aquat. Microb. Ecol.* 42:41-53.

Hoegh-Guldberg, O. (2011). Coral reef ecosystems and anthropogenic climate change. *Reg. Environ. Change* 11(Suppl 1):215-227.

Hoffmann, J. P. (1998). Wastewater treatment with suspended and non-suspended algae. *J. Phycol.* 34:757-763.

Holligan, P. M. & Robertson, J. E. (1996). Significance of ocean carbon budgets for the global carbon cycle. *Global Change Biol.* 2:85-95.

Hong, D. D., Hoang, M. H., Hoang, S. N., Nguyen, D. H., Hoang, L. A., Ngo, H. T. & Dinh, K. C. (2007). Biodiversity of heterotrophic microalgal *Labyrinthula* sp. and *Schizochytrium* sp. of Vietnam. In; Con, V. Q. (Ed.). *Proceedings of the 2nd National Scientific Conference on Ecology and Biological Resources, Hanoi.* Agriculture Publishing House, Hanoi, Vietnam, pp332-338. (in Vietnamese).

Hong, D. D., Anh, H. T. L. & Thu, N. T. H. (2011). Study on biological characteristics of heterotrophic marine microalga – *Schizochytrium mangrovei* PQ6 isolated from Phu Quoc Island, Kien Giang province, Vietnam. *J. Phycol.* 47:944-954.

Horton, P., Ruban, A. V. & Walters, R. G. (1996). Regulation of light harvesting in green plants. *Ann. Rev. Plant Physiol and Plant Mol. Biol.* 47:655-684.

Houchins, J. P. (1984). The physiology and biochemistry of hydrogen metabolism in cyanobacteria. *Biochim. Biophys. Acta* 768:227-255.

Houghton, J. T., Ding, Y., Griggs, D. J., Noguer, M., van der Winden, P. J. & Dai, X. (2001). Climate change 2001: the scientific basis. Contribution of Working Group I to the Third Assessment Report of the International Panel on Climate Change. International Panel on Climate Change, Cambridge University Press, Cambridge.

Hu, Q., Sommerfield, M., Jarvis, E., Ghirardi, M., Posewitz. M., Seibert, M. & Darzins, A. (2008). Microalgal triacylglycerols as feedstocks for biofuel production: perspectives and advances. *Plant J.* 54:621-639.

Huang, C-C., Chen, M-W., Hsieh, L-J., Lin, W-H., Chen, P-C. & Chien, L-F. (2006). Expression of mercuric reductase from *Bacillus megaterium* MB1 in eukaryotic microalga *Chlorella* sp. DT: an approach for mercury phytoremediation. *Appl. Microbial Biotechnol.* 72:197-205.

Huang, J-C., Wang, Y., Sandmann, G. & Chen, F. (2006). Isolation and characterization of a carotenoid oxygenase gene from *Chlorella zofingensis* (Chlorophyta). *Appl. Microbiol. Biotechnol.* 71:473-479.

Huber, V., Adrian, R. & Gerten, D. (2008). Phytoplankton response to climate warming modified by trophic state. *Limnol. Oceanogr.* 53:1-13.

Humpage, A. R. & Falconer, I. R. (2003). Oral toxicity of the cyanobacterial toxin cylindrospermopsin in male Swiss albino mice: Determination of no observable adverse effect level for deriving a drinking water quideline value. *Environmental Toxicology* 18:94-103.

Hurd, C. L., Hepburn, C. D., Currie, K. I., Raven, J. A. & Hunter, K. A. (2009). Testing the effects of ocean acidification on algal metabolism: considerations for experimental designs. *J. Phycol.* 45:1236-1251.

Hutadilok-Towatana, N., Reanmongkol, W. & Panichayupakaranant, P. (2010). Evaluation of the toxicity of *Arthrospira* (*Spirulina*) *platensis* extract. *J. Appl. Phycol.* 22:599-605.

Hutchins, D. A., Fu, F. X., Zhang, Y., Warner, M. E. and others. (2007). CO_2 control of *Trichodesmium* N_2 fixation, photosynthesis, growth rates, and elemental ratios: implications for past, present, and future ocean biogeochemistry. *Limnol. Oceanogr.* 52:1293-1304.

Ikuta, Y., Akano, T., Shioji, N. & Maeda, I. (1998). Hydrogen Production by Photosynthetic Microorganisms. In Zaborsky, O. (ed.), *Biohydrogen,* Plenum Press, New York, pp. 319-328.

Ichimi, K., Kawamura, T., Yamamoto, A., Tada, K. & Harrison, P. J. (2012). Extremely high growth rate of the small diatom *Chaetoceros salsugineum* isolated from an estuary in the eastern seto inland sea, Japan. *J. Phycol.* 48:1284-1288.

Imamul-Hug, S. M., Abdullah, M. B. & Joardar, J. C. (2007). Bioremediation of arsenic toxicity by algae in rice culture. *Land Contamination & Reclamation* 15:327-334.

Innok, S., Chunleuchanon, S., Boonkerd, N. & Teaumroong, N. (2009). Cyanobacterial akinete induction and its application as biofertilizer for rice cultivation. *J. Appl. Phycol.* 21:737-744.

Ip, P-F. & Chen, F. (2005). Production of astaxanthin by the green microalga *Chlorella zofingensis* in the dark. *Process Biochemistry* 40:733-738.

Ito, T., Kurano, N., Soga, T. & Tomita, M. (2009). Lipid-metabolite analysis of oil-rich alga, *Pseudochoricystis ellipsoidea. Phycologia,* 48. Abstract no. 142, 9th Internatl. Phycol. Congress, Tokyo.

Izaguirre, G, Jungblut, A-D. & Neilan, B. A. (2007). Benthic cyanobacteria (Oscillatoriaceae) that produce microcystin-LR, isolated from four reservoirs in southern California. *Water Research* 41:492-498.

Jacobsen, A., Grahl-Nielsen, O. & Magnesen, T. (2010). Does a large-scale continuous algal production system provide a stable supply of fatty acids to bivalve hatcheries? *J. Appl. Phycol.* 22:769-777.

Jacobsen, A., Grahl-Nielsen, O. & Magnesen, T. (2012). Effects of reduced diameter of bag cultures on content of essential fatty acids and cell density in a continuous algal production system. *J. Appl. Phycol.* 24:109-116.

Jans, F and others (2008). A type II NAD(P)H dehydrogenase mediates light-independent plastoquinone reduction in the chloroplast of *Chlamydomonas. Porc. Natl. Acad. Sci. USA* 105:20546-20551.

Jansson, C. (2012). Metabolic engineering of cyanobacteria for direct conversion of CO_2 to hydrocarbon biofuels. In: Lüttge et al. (eds.), *Progress in Botany* 73 pp81-93. Springer-Verlag, Berlin, Heidelberg.

Jasso-Chávez, R., Pacheco-Rodales, A., Lira-Silva, E., Gallardo-Pérez, J. C., García, N. & Moreno-Sánchez, R, (2010). Toxic effects of Cr(VI) and Cr(III) on energy metabolism of heterotrophic *Euglena gracilis. Aquatic Toxicology* 100:329-338.

Jeffrey, S. W., MacTavish, H. S., Dunlap, W. C., Vesk, M. & Groenewoud, K. (1999). Occurrence of UVA- and UVB-absorbing compounds in 152 species (206 strains) of marine microalgae. *Mar. Ecol. Prog. Ser.* 189:35-51.

Jensen, S. & Jernelõv, A. (1969). Biological methylation of mercury in aquatic organisms. *Nature,* 223:753-754.

Jiang, H. & Qui, B. (2011). Inhibition of photosynthesis by UV-B exposure and its repair in the bloom-forming cyanobacterium *Microcystis aeruginosa. J. Appl. Phycol.* 23:691-696.

Jodlowska, S. & Latala, A. (2010). Photoacclimation strategies in the toxic cyanobacterium *Nodularia spumigena* (Nostocales, Cyanobacteria). *Phycologia* 49:203-211.

Johnson, P. W. & Sieburth, J. McN. (1979). Chrooccoid cyanobacteria in the sea: A ubiquitous and diverse phototrophic biomass. *Limnol. Oceanogr. 24:928-935.*

Johnston, A. M., Raven, J. A., Beardall J. & Leegood, R. C. (2001). Photosynthesis in a marine diatom. *Nature* 412:40-41.

Johnsen, T. M. & Lien, T. E. (1989). *Prymnesium parvum* Carter (Prymnesiophyceae) in association with macroalgae in Ryfylke, Southwestern Norway. *Sarsia* 74:277-281.

Jones, G. J. & Korth, W. (1995). *In situ* production of volatile odour compounds by river and reservoir phytoplankton populations in Australia. *Water Science & Technology* 31:145-157.

Jones, H. L. J., Leadbeater, B. S. C. & Green, J. C. (1994). Mixotrophy in haptophytes. In: J. C. Green & B. S C. Leadbeater (eds.). *The Haptophyte Algae* pp. 247-264.

Jordan, M. A. & Wilson, L. (1998). Microtubules and actin filaments: dynamic targets for cancer chemotherapy. *Curr. Opin. Cell Biol.* 10:123-130.

Kaczmarzyk, D. & Fulda, M. (2010). Fatty acid activation in cyanobacteria mediated by acyl-acyl carrier protein synthetase enables fatty acid recycling. *Plant Physiol.* 152:1598-1610.

Kaiho, K (1994). Planktonic and benthic foraminferal extinction events during the last 100my. *Planktonic, Palaeoclimatology, Palaeoecology* 111:45-71.

Kannaiyan, S. (1993). Nitrogen contribution by *Azolla* to rice crops. *Proc.Indian Nat. Sci. Acad.* B59:309-314.

Kannaiyan, S., Aruna, S. J., Kumari, M. P. & Hall, D. O. (1997). Immobilized cyanobacteria as a biofertilizer for rice crops. *J. Appl. Phycol.* 9:167-174.

Karentz, D., McEuen, F. S., Land, M. C. & Dunlap, W. C. (1991). Survey of mycosporine-like amino acid compounds in Antarctic marine organisms: potential protection from ultraviolet exposure. *Mar. Biol.* 108:157-166.

Karlsson, A., Auer, N., Schulz-Bull, D. & Abrahamsson, K. (2008). Cyanobacterial blooms in the Baltic – A source of halocarbons. *Marine Chemistry* 110:129-139.

Kaya, V. M., De la Noüe, J. & Picard, G. (1985). A comparative study of four systems for for tertiary wastewater treatment by *Scenedesmus bicellularis*: New technology for immobilization. *J. Appl. Phycol.* 7:85-95.

Kebede-Westhead, E., Pizarro, C., Mulbry, W. W. & Wilkie, A. C. (2003). Production and nutrient removal by periphyton grown under different loading rates of anaerobically digested flushed dairy manure. *J. Phycol.* 39:1275-1282.

Kehrer, J. P. & Smith, C. S. (1994). Free radicals in biology: sources, reactivities and roles in the etiology of human diseases. In: Frei, B (ed) *Natural Antioxidants in Human Health and Disease.* Academic Press, San Diego, pp 25-62.

Kelly, A. M., Holmes, W. E. & Schultz, T. P. (2006). An environmentally benign method for reducing off-flavour terpenoids in aquaculture: Initial laboratory studies. *North American J. Aquaculture* 68:249-252.

Kennedy, S. B., Tucker, J. W., Neidig, C. L.,Vermeer, G. K., Cooper, V. R., Jarrell, J. L. & Sennet, D. G. (1998). Bacterial management strategies for stock enhancement of warmwater marine fish: a case study with common snook (*Centropomus undecimalis*). *Bull. Mar. Sci.* 62:573-588.

Kesarcodi-Watson, A., Klumpp, D. W. & Lucas, J. S. (2001). Comparative feeding and physiological energetics in diploid and triploid Sydney rock oysters (*Saccostrea commercialis*). II. Influences of food concentration and tissue energy distribution. *Aquaculture* 203:195-216.

Kessler, E. (1956). Reduction of nitrite with molecular hydrogen in algae containing hydrogenase. *Arch. Biochem. Biophys.* 62:241-242.

Khan, Z., Bhadouria, P. & Bisen, P. S. (2005). Nutritional and therapeutic potential of *Spirulina. Current Pharmaceutical Biotechnology,*6:373-379.

Kharrat, R., Servenr, D., Girard, E., Ouanounou, G., Amar, M., Marrouchi, R., Benoit, E. & Molgó, J. (2008). The marine phycotoxin gymnodimine targets muscular and neuronal nicotinic acetylcholine receptor subtypes with high affinity. *J. Neurochem.* 107:952-963.

Khursheed, S., Anwer, R., Zutshi, S. & Fatma, T. (2012). Screening of photosynthetic O_2 – evolving prokaryotes for an insulin-like antigen. *J. Phycol.* 48:243-245.

Kiene, R. P. & Linn, L. J. (2000). Distribution and turnover of dissolved DMSP and its relationship with bacterial production and dimethyl sulphide in the Gulf of Mexico. *Limnol. Oceanogr.* 45:849-861.

Kim, E-K., Choi, G-G., Kim, H-S., Ahn, C-Y & Oh, H-M. (2012). Increasing γ-linolenic acid content in *Spirulina platensis* using fatty acid supplement and light-dark illumination. *J. Appl. Phycol.* 24:743-750.

Kirkpatrick, B., Fleming, L. E., Squiccarini, D., Backer, L. C. and others (2004) Literature review of Florida red tide: implications for human health. *Harmful Algae* 3:99-115.

Klein, U. & Betz, A. (1978). Fermentive metabolism of hydrogen-evolving *Chlamydomonas moewusii. Plant Physiol.* 61:953-956.

Kliphuis, A. M., de Winter, L., Vejraska, C., Martens, D. E., Janssen, M. & Wijffels, R. H. (2010). Photosynthetic efficiency of *Chlorella sorokiniana* in a turbulent mixed short light-path photobioreactor. *Biotechnol. Prog.* 26:687-696.

Klisch, M. & Häder, D-P. (2000). Mycosporine-like amino acids in the marine dinoflagellate *Gyrodinium dirsum*: induction by ultraviolet irradiation. *J. Photochem. Photobiol* B 55:178-182.

Klisch, M. & Häder, D-P. (2002). Wavelength dependence of mycosporine-like amino acid synthesis in *Gyrodinium dorsum. J. Photochem. Photobiol* B 66:60-66.

Knaff, D. B. (1996). Ferredoxin and ferredoxin-dependent enzymes. *In* Oxygenic Photosynthesis: The Light Reactions. D. R. Ort and C. F. Yocum (Eds.) Kluwer Academic Publishers, Dordrecht, The Netherlands.

Knauer, K., Behra, R. & Hemond, H. (1999). Toxicity of inorganic and methylated arsenic to algal communities from lakes along an arsenic contamination gradient. *Aquat. Toxicol.* 46:221-230.

Knights, B. A., Brown, A. C., Conway, E. & Middleditch, B. S. (1970). Hydrocarbons from the green form of the of the freshwater alga *Botryococcus braunii. Phytochemistry* 9:1317-1324.

Knothe, G. (2005). Dependence of biodiesel fuel properties on the structure of fatty acid alkyl esters. *Fuel Proc. Technol.* 86:1059-1070.

Knothe, G. (2009). Improving biodiesel fuel properties by modifying fatty ester composition. *Energ. Environ. Sci.* 2:759-766.

Kobayashi, M. (2000). *In vivo* antioxidant role of astaxanthin uner oxidative stress in the green alga *Haematococcus pluvialis. Appl. Microbial Biotechnol.* 54:550-555.

Kodama, M., Ikemoto, H. & Miyachi, S. (1993). A new species of highly CO_2-tolerant fast growing marine microalga for high-density cultivation. *J. Mar. Biotech.*1:21-25.

Koistra, W., Gersonde, R., Medlin, L. & Mann, D. G. (2007). In: Falkowski, P. G. & Knoll, A. H. (eds.) *Evolution of Primary Producers in the Sea* pp207-249. Academic Press, New York.

Koo, J., Bai, S. C., Kim, K. & Kim, S. (2001). Optimum dietary level of *Chlorella* powder as a feed additive for growth performance of juvenile olive flounder, *Paralichthys olivaceus. J. Applied Aquaculture* 11:55-66.

Koo, S. Y., Cha, K. H., Song, D-G., Chung, D. & Pan, C-H. (2012). Optimization of pressurized liquid extraction of zeaxanthin from *Chlorella ellipsoidea. J. Appl. Phycol.* 24:725-730.

Kosourov, S., Siebert, M & Ghirardi, M. L. (2003). Effects of extracellular pH on the metabolic pathways in sulphur-deprived H_2 –producing *Chlamydomonas reinhardtii. Plant Cell Physiol.* 44:146-155.

Krinsky, N. I. & Johnson, E. J. (2005) Carotenoid actions and their relation to health and disease. *Mol. Asp. Med.* 26:459-516.

Kroth, P. (2007). Molecular biology and the biotechnological potential of diatoms. Springer, Berlin.

Kruse, O., Rupprecht, J., Bader, K. P., Thomas-Hall, S., Schenk, P. M., Finazzi, G. & Hankamer, B. (2005). Improved photobiological H_2 production in engineered green algal cells. *J. Biol. Chem.* 280:34170-34177.

Kudela, R. M. (2010). Does horizontal mixing explain phytoplankton dynamics? *Proc. Nat. Acad. Sci. USA* 107:18235-18236.

Kumar, R. & Goyal, D. (2010). Waste water treatment and metal (Pb^{2+}, Zn^{2+}) removal by microalgal based stabilization pond system. *Indian J. Microbiol.* 50 (suppl 1):S34-S40.

Kurano, N. (2009). CO_2 removal by microalgae. 9th Int. Phycol. Congr. Tokyo. *Phycologia* 48. Abstract No. 199. p.69.

Kurano, N. & Miyachi, S. (2004). Microalgal studies for the 21st Century. *Hydrobiologia* 512:27-32.

Kurmayer, R. (2011). The toxic cyanobacterium *Nostoc* sp. strain 152 produces highest amounts of microcystin and nostophycin under stress conditions. *J. Phycol.* 47:200-207.

Kuznetsova, L. A. & Coakley, W. T. (2007). Applications of ultrasound streaming and radiation force in biosensors. *Biosensors & Bioelectronics* 22:1567-1577.

Kyle, D. J., Reeb, S.E. & Sicotte, V. J. (1998). Dinoflagellate biomass, methods for its production, and compositions containing the same. USA Patent 5,711,983.

Laguna, M. R., Villar, R., Calleja, J. M. & Cadavid, I. (1993a). Effects of *Chlorella stigmatophora* extracts on the central nervous system. *Planta Med.* 59:125-130.

Laguna, M. R., Villar, R., Cadavid, I. & Calleja, J. M. (1993b). Effects of extracts of *Tetraselmis suecica* and *Isochrysis galbana* on the central nervous system. *Planta Med.* 59:207-214.

Lampert W. (1978). Release of dissolved organic carbon by grazing phytoplankton. *Limnol. Oceanogr.* 23:831-834.

Lasker, R. & Zweifel, J. R. (1978). Growth and survival of first-feeding northern anchovy larvae (*Engraulis mordax*) in patches containing different proportions of large and small prey. In Steele, J. H. (Ed.) *Spatial patterns in plankton communities.* Pp. 329-354. Plenum Press, New York.

Last, J. M. (1980). The food of twenty species of fish larvae in the west-central North Sea. *Fish. Res. Tech. Rep., MAFF Direct. Fish Res., Lowestoft* 60:1-44.

Laurinavichene, T. V., Kosourov, S. N., Ghirardi, M. L., Seibert, M. & Tsygankov, A. A. (2008). Prolongation of H_2 photoproduction by immobilized, sulphur-limited *Chlamydomonas reinhardtii* cultures. *J. Biotechnol.* 134:275-277.

Leaf, A. & Weber, P. C. (1988). Medical progress. Cardiovascular effects of n-6 fatty acids. *N. Engl. J. Med.* 318:549-557.

Leão, P. N. and others. (2010). Synergistic allelochemicals from a freshwater cyanobacterium. *Proc. Natl. Acad. Sci.USA* 107:11183-11188.

Lebeau, T. & Robert, J. M. (2003). Diatom cultivation and biotechnologically relevant products. Part I: cultivation at various scales. *Appl. Microbiol. Biotechnol.* 60:612-623.

Leckey, C. A. C. & Hinders, M. K. (2011). Viscous effects in the acoustic manipulation of algae for biofuel peoduction. *J. Appl. Phycol.* DOI 10.1007/s10811-011-9662-7.

Lee, J. S., Kim, D. K., Lee, J. P., Park, S. C., Koh, H. S., Cho, H. S. & Kim, S. W. (2002). Effects of SO_2 and NO on growth of *Chlorella* sp.KR-1. *Biores. Biotechnol.* 82:1-4.

Lee, Y-K. (2001). Microalgal mass culture systems and methods: Their limitation and potential. *J. Appl. Phycol.* 13:307-315.

Lee, P. D., Deonarine, A. S., Kienetz, M and others. (2001). A novel pathway for lipid biosynthesis: the direct acylation of glycerol. *J. Lipid Res.*. 42:1979-1986.

Lee, Y. K. (1997). Commercial production of microalgae in the Asia-Pacific rim. *J. Appl. Phycol.* 9:403-411.

Lee, A. K., Lewis, D. M. & Ashman, P. J. (2009). Microbial flocculation, a potentially low-cost harvesting technique for marine microalgae for the production of biodiesel. *J. Appl. Phycol.* 21:559-567.

Legendre, L. & Le Fèvre, J. (1995). Microbial food webs and the export of biogenic carbon in oceans. *Aquat. Microbial Ecol.* 9:69-77.

Legendre, L. & Michaud, J. (1998). Flux of biogenic carbon in oceans: size-dependent regulation by pelagic food webs. *Mar. Ecol. Prog. Ser.* 164:1-11.

Lei, A. P., Wong, Y. S. & Tam, N. F. Y. (2002). Removal of pyrene by different microalgal species. *Water Sci. Technol.* 46:195-201.

Leira, M. (2005). Diatom responses to Holocene environmental changes in a small lake in northwestern Spain. *Quaternary International* 140-141:90-120.

Lelong, A., Hégaret, H., Soudant, P. & Bates, S. (2012). *Pseudo-nitzschia* (Bacillariophyceae) species, domoic acid and amnesic shellfish poisoning: revisiting previous paradigms. *Phycologia* 51:168-216.

Levitus, S., Antonov, J. I., Boyer, T. P. & Stephens, C. (2000). Warming of the world ocean. *Science* 287:2225-2229.

Li, D. M. & Qi, Y. Z. (1997). *Spirulina* industry in China: present status and future prospects. *J. Appl. Phycol.* 9:25-28.

Li, Y., Horsman, M., Wang, B.,Wu, N & Lan, C. Q. (2008). Effects of nitrogen sources on cell growth and lipid accumulation of green alga *Neochloris oleoabundans. Appl. Microbiol. Biotechnol.* 81:629-636.

Li, Y., Sommerffeld, M., Chen, F. & Hu, Q. (2010). Effect of photon flux densities on regulation of carotenogenesis and cell viability of *Haematococcus pluvialis* (Chlorophyceae). *J. Appl. Phycol.* 22:253-263.

Li, Z., Keasling, J. D. & Niyogi, K. K. (2012). Overlapping photoprotective function of vitamin E and carotenoids in *Chlamydomonas. Plant Physiol.* 158:313-323.

Liang, S., Liu, X., Chen, F. & Chen, Z. (2004). Current microalgal health food R & D activities in China. *Hydrbiologia* 512:45-48.

Lim, P-T., Leaw, C-P. & Usup, G. (2006). Effects of light and temperature on growth, nitrate uptake and toxin production of two tropical dinoflagellates: *Alexandrium tamiyavanichii* and *Alexandrium minutum* (Dinophyceae). *J. Phycol.* 42:786-799.

Lindell, D. & Post, A. F. (1995). Ultraphytoplankton succession is triggered by deep winter mixing in the Gulf of Aqaba (Eilat), Red Sea. *Limnol. Oceanogr.* 40:1130-1141.

Link, J. S., Gaichas, S., Miller, T. J. and others. (2012). Synthesizing lessons learned from comparing fisheries production in 13 northern hemisphere ecosystems: emergent fundamental features. *Mar. Ecol. Prog. Ser.* 459:293-302.

Lionard, M., Roy, S., Tremblay-Létourneau, M. & Ferreyra, G. A. (2012). Combined effects of increased UV-B and temperature on the pigment-determined marine phytoplankton community of the St. Lawrence Estuary. *Mar. Ecol. Prog. Ser.* 445:219-234.

Lipps, J. H. (1970). Plankton evolution. *Evolution* 24:1-22.

Liss, P. S., Malin, G. & Turner, S. M. (1992). Production of DMS by marine phytoplankton. In: Restelli, G. & Angelethi, G. (eds.) *Dimethylsulphide: Oceans, Atmosphere & Climate.* Proc. Int. Symposium. Belgerate, Italy, Oct. 1992. pp.1-14. Kluwer Academic Publishers, Dordrecht, The Netherlands.

Livingstone. D. M. & Dokulil, M. T. (2001). Eighty years of spatially coherent Austrian lake surface temperatures and their relationship to regional air temperatures and the North Atlantic Oscillation. *Limnol. Oceanogr.* 46:1220-1227.

Llama, M., Serra, J. L., Rao, K. K. & Hall, D. O. (1981). Isolation of two hydrogenase activities in *Chromatium. Eur. J. Biochem.* 114:89-96.

Logan, B. E., Passow, U., Alldredge, A. L., Grossart, H. P. & Simon, M. (1995). Rapid formation and sedimentation of large aggregates is predictable from coagulation rates (half-lives) of transparent exopolymer particles (TEP). *Deep-Sea Research II* 42:203-214.

Lomas, M. W., Hopkinson, B. M., Losh, J. L. and others (2012). Effect of ocean acidification on cyanobacteria in the subtropical North Atlantic. *Aquat. Microb. Ecol.* 66:211-222.

Long, B. M., Tucker, L., Badger, M. R.& Price, G. D. (2010). Functional cyanobacterial beta-carboxysomes have an absolute requirement for both long and short forms of the CCM protein. *Plant Physiol.* 153:285-293.

López-Rodas, V., Agrelo, M., Carillo, E., Ferrero, L.M., Larrauri, A., Martin-Otero, L. & Costas, E. (2001). Resistance of microalgae to modern water contaminants as the result of rare spontaneous mutation. *Eur. J. Phycol.* 36:179-190.

Lorenz, R. T. & Cysewski, G. R. (2000). Commercial potential for *Haematococcus* microalgae as a natural source of astaxanthin. *Trends in Biotechnol.* 18:160-167.

Lorenzo, J. I., Nieto-Cid, M., Álvarez-Salgado, X. A., Pérez, P. & Beiras, R. (2007). Contrasting complexing capacity of dissolved organic matter produced during the onset, development and decay of a simulated bloom of the marine diatom *Skeletonema costatum. Marine Chemistry* 103:61-75.

Lu, Y., Ding, Y. & Wu, Q. (2011). Simultaneous saccharification of cassava starch and fermentation of algae for biodiesel production. *J. Appl. Phycol.* 23:115-121.

Lu, Y-M., Xiang, W-Z. & Wen, Y-H. (2011). *Spirulina (Arthrospira)* industry in Inner Mongolia of China: current status and prospects. *J. Appl. Phycol.* 23:265-269.

Lubián, L. M. & Establier, R. (1982). Comparative study of the pigment composition in several strains of *Nannochloropsis* (Eustigmatophyceae) during growth in culture. *Inv. Pesq.* 46:379-389.

Lubián, L. M., Establier, R. & Ruis, E. (1986). Light and nitrate influence on the growth and pigment contents of *Nannochloropsis gaditana* Lubián in culture. *Inv. Pesq.* 50:499-518.

Lubián, L. M., Montero, O., Moreno-Garrido, I., Huertas, I. E., Sobrino, C., Valle, M. G-del. & Parés, G. (2000). *Nannochloropsis* (Eustigmatophyceae) as source of commercially valuable pigments. *J. Appl. Phycol.* 12:249-255.

Lubzens, E., Marko, A. & Tietz, A. (1985). *De novo* synthesis of fatty acids in the rotifer, B*rachionus plicatilis*. *Aquaculture* 47:27-37.

Lumpkin, T. A. & Plunckett, D. L. (1980). *Azolla*: Botany, physiology, and use as a green manure. *Economic Botany* 34:111-153.

Maas, E. W. & Brooks, H. J. L. (2010). Is photosynthesis a requirement for parlytic shellfish toxin production in the dinoflagellate *Alexandrium minutum* algal-bacterial consortium? *J. Appl. Phycol.* 22:293-296.

Madronich, S., McKenzie, R. L., Bjorn, L. O. & Calewell, M. M. (1998). Changes in biologically active ultraviolet radiation reaching the Earth's surface. *J. Photochem. Photobiol. B. Biol.* 46:5-19.

Maeda, K., Owada, M., Kimura, N., Omata, K. & Karube, I. (1995). CO_2 fixation from the flue gas of coal-fired thermal power plant by microalgae. *Energy Conservation Management* 36:717-720.

Magnuson, J. J. and others. (2003) Historical trends in lake and river ice cover in the Northern Hemisphere. *Science* 289:1743-1746.

Mague, T. H., Friberg, E., Hughes, D. J. & Morris, I. (1980). Extracellular release of carbon by marine phytoplankton – a physiological approach. *Limnol. Oceanogr.*25:262-279.

Maione, T. E. & Gibbs, M. (1986). Hydrogenase-mediated activities in isolated chloroplasts of *Chlamydomonas reinhardtii. Plant Physiol.* 80:360-363.

Malin, G. & Grimalt, J. O. (2000). Biologocal turnover of DMS, DMSP, and DMSO in contrasting open-sea waters. *Mar.Ecol. Prog. Ser.* 203:1-11.

Malin, G., Wilson, W. H., Bratbak, G, Liss, P. S. & Mann, N. H. (1998). Elevated production of dimethylsulfide resulting from viral infection of cultures of *Phaeocystis pouchetti. Limnol. Oceanogr.* 43:1389-1393.

Malin, G. & Kirst, G. O. (1997). Algal production of dimethyl sulphide and its atmospheric role. *J. Phycol.* 33:889-896.

Malin, G., Liss, P. S. & Turner, S. M. (1994). Dimethyl sulphide: production and atmospheric consequences. In: Green, J. C. & Leadbeater, B. S. C. (eds.) *The Haptophyte Algae.* Clarendon Press, Oxford.

Malin, G. & Steinke, M. (2004). Dimethyl sulphide production: what is the contribution of the coccolithophores? In: Thierstein, H. R. & Young, J. R. (eds.) *Coccolithophores: From molecular processes to global impact.* Springer-Verlag, Berlin, Heidelberg.

Mandal, S. & Mallick, N. (2009). Microalga *Scenedesmus obliquus* as a potential source for biodiesel production. *Appl. Microbiol. Biotechnol.* 84:281-291.

Manuel, C. D. Jr. & Neyra-Tanabe. (2012). Mercury uptake of microalgae in Kematu River, Teboli, South Corabuto, Philippines. Special Publication, University of Southern Mindanao, Philippines.

Margalith, P. Z. (1999). Production of ketocarotenoids by microalgae. *Appl. Microbiol. Biotechnol.* 51:431-438.

Mari, X. & Kiørboe, T. (1996). Abundance, size distribution and bacterial colonisation of transparent exopolymer particles (TEP) during spring in the Kattegat, *J. Plankton Res.* 18:969-986.

Martin, J. F., Izaguirre, G. Waterstrat, P. (1991). A planktonic *Oscillatoria* species from Mississippi catfish ponds that produces the off-flavour compound 2-methylisoborneol. *WaterRes.* 25:1447-1451.

Martiny, A. C., Kathuria, S. & Berube, P. M. (2009). Widespread metabolic potential for nitrite and nitrate assimilation among *Prochlorococcus* ecotypes. *Proc. Nat. Acad. Sci.* 106:10787-10792.

Marxen, K., Vanselow, K. H., Hintze, R., Lippemeier, S., Ruser, A., Egge, B., Colijin, F. & Hansen, U-P. (2010). Comparison of two different modes of UV-B irradiation on synthesis of some cellular substances in the cyanobacterium *Synechocystis* sp. PCC6803. *J. Appl. Phycol.* 22:677-690.

Masojídek, J., Kopecký, J., Koblíek, M. & Torzillo, G. (2004). The xanthophyll cycle in green algae (Chlorophyta): its role in the photosynthetic apparatus. *Plant Biology* 6:342-349.

Masojídek, J., Sergejevová, M., Rottnerová, K., Jirka, V., Korečko, J., Kopecký, J., Zat'ková, I., Torzillo, G. & Štys, D. (2009). A two-stage solar photobioreactor for cultivation of microalgae based on solar concentrators. *J. Appl. Phycol.* 21:55-63.

Mason, C. P. et al. (1982). Isolation of chlorine-containing antibiotic from the freshwater cyanobacterium *Scytonema hofmannii. Science* 213:400-402.

Masukawa, H., Mochimaru, M. & Sakurai, H. (1998). Some characteristics of regulation of nitrogenase and hydrogen producing activities in cyanobacteria. In: Garab, G. (ed.), *Photosynthesis: Mechanisms and Effects Vol. V, 4143-4146.* Kluwer Academic, Netherlands.

Marsui, M. S., Muizzuddin, N., Arad, S. & Marens, K. (2003). Sulfated polysaccharides from red microalgae have antiinflammatory properties *in vitro* and *in vivo. Appl. Biochem. Biotechnol.* 104:13-22.

Matsuda, Y., Hara, T. & Colman, B. (2001). Regulation of the induction of bicarbonate uptake by dissolved CO_2 in the marine diatom *Phaeodactylum tricornutum. Plant Cell & Environment* 24:611-620.

Matsunga, S., Moore, R. E., Niemczura, W. P. & Carmichael, W. W. (1989). Anatoxin-a(s), a potent anticholinesterase from *Anabaena flos-aquae. J. Am. Chem. Soc.* 111:8021-8023.

Maxwell, J. R., Douglas, A. G., Eglington, G. & McCormick, A. (1968). The botryococcenes - hydrocarbons of novel structure from the alga *Botryococcus braunii,* Kützing. *Phytochemistry* 7:2157- 2171.

McCarty, M. F. (2007). Clinical potential of *Spirulina* as a source of Phycocyanobilin. *J. Mwedicinal Food.* 10:566-570.

McConnaughey, T. A. (2012). Zooxanthellae that open calcium channels: implications for reef corals. *Mar. Ecol. Prog. Ser.* 460:277-287.

McKnight, D., Brakke, D. F. & Mulholland, P. J. (1996). Freshwater ecosystems and climate change in North America. *Limnol. Oceanogr.* 41:815-1149.

McQuatters-Gollop, A., Raitsos, D. E.. Edwards, M. & Attrill, M. J. (2007). Spatial patterns of diatom and dinoflagellate seasonal cycles in the NE Atlantic Ocean. *Mar. Ecol. Prog. Ser.* 339:301-306.

Medlin, L. K. (2011). The Permian-Triassic mass extinction forces the radiation of the modern marine phytoplankton. *Phycologia* 50:684-693.

Meiser, A., Schmid-Staiger, U. & Trösch, W. (2004). Optimization of eicosapentaenoic acid production by *Phaeodactylum tricornutum* in the flat panel airlift (FPA) reactor. *J. Appl. Phycol.* 16:215-225.

Melis, A. (2009). Solar energy conversion efficiencies in photosynthesis: Minimizing the chlorophyll antennae to maximize efficiency. *Plant Science* 177:272-280.

Melis, A. & Chen, H-C. (2005). Chloroplast sulfate transport in green algae, genes proteins and effects. *Photosynth Res.* 86:299-300.

Melis, A., Zhang, L. P., Forestier, M., Ghirardi, M. L. & Siebert, M. (2000). Sustained photobiological hydrogen gas production upon reversible inactivation of oxygen evolution in the green alga *Chlamydomonas reinhardtii. Plant Physiol.* 122:127-135.

Merico, A., Tyrrell, T. & Cokacar, T. (2006). Is there any relationship between phytoplankton seasonal dynamics and the carbonate system? *J. Mar. Syst.* 59:120-142.

Metcalf, J. S., Lindsay, J., Beattie, K. A., Birmingham, S., Saker, M. L., Törökné, A. K. & Codd, G. A. (2002). Toxicity of cylindrospermopsin to the brine shrimp *Artemia salina*: comparisons with protein synthesis inhibitors and microcystins. *Toxicon* 40:1115-1120.

Metting, B. (1988). Micro-algae in agriculture. In: Borowitzka, M. A. & Borowitzka, L. J. (Eds.). *Micro-algal Biotechnology* pp.288-303. Canbridge Univ. Press, Cambridge, England.

Metting, B. & Rayburn, W. R. (1983). The influence of a microalgal conditioner on selected Washington soils: an empirical study. *Soil Science Society of America Journal,* 47:682-685.

Metz, B., Davidson, O. R., Bosch, P. B., Dave, R. & Meyer, L. A. (eds.) *Contribution to the Fourth Assessment Report of the International Panel on Climate Change,2007. Chapter 7.4.5.1 "Cement".* Resource document http://www.ipcc.ch/publications and data/ar4/wg3/en/ch7-ens7-4-5.html.

Metzger, P. & Largeau, C. (2005). *Botryococcus braunii*: a rich source for hydrocarbons and related ether lipids. *Appl. Microbiol. Biotechnol.* 66:486-496.

Miao, A. J., Schwehr, K. A., Xu, C.,Zhang, S. J., Luo, Z., Quigg, A. & Santschi, P. H. (2009). The algal toxicity of silver engineered nanoparticles and detoxification by exopolymeric substances. *Environ. Pollut.* 157:3034-3041.

Miao, X. L. & Wu, Q. Y. (2004). High yield bio-oil from fast pyrolysis by metabolic controlling of *Chlorella protothecoides. J. Biotechnol.*110:85-93.

Miao, X. L. & Wu, Q. Y. (2006). Biodiesel production from heterotrophic microalgal oil. *Bioresource Technol.* 97:841-846.

Middleburg,J. J. & Meysman, F. J. R. (2007). Burial at sea. *Science* 317:1294-1295.

Milliman, J. D. (1993). Production and accumulation of calcium carbonate in the ocean: budget of an unsteady state. *Global Biogeochemical Cycles.* 7:927-957.

Minowa, T., Yokoyama, S-Y., Kishimoto, M. & Okakurat, T. (1995). Oil production from algal cells of *Dunaliella tertiolecta* by direct thermochemical liquefaction. *Fuel* 74:1735-1738.

Mishra, U. & Pabbi, S. (2004). Cyanobacteria: a potential biofertilizer for rice. *Resonance* 9:6-10.

Mitsuhashi, S., Mizushima, T., Yamashita, E. and others (2000). X-ray structure of β-carbonic anhydrase from the red alga, *Porphyridium purpureum,* reveals a novel catalytic site for CO_2 hydration. *J. Biol. Chem.* 275:5521-5526.

Miyamoto, K., Ohta, S., Nawa, Y., Mori, Y & Miura, Y. (1987). Hydrogen production by a mixed culture of a grren alga, *Chlamydomonas reinhardtii* and a photosynthetic bacterium *Rhodospirillum rubrum. Agr. Biol. Chem.* 51:1319-1324.

Mohammed, M. H. & Markert, B. (2006). Toxicity of heavy metals on *Scenedesmus quadricauda* (Turp.) de Brébisson in batch cultures. *Environ. Sci. Pollut. Res.* 13:98-104.

Mohan, M. & Mukerji, K. G. (1978). Some biologically active extracellular products of blue-green algae. *Phykos,*18:73-82.

Moheimani, N. R. (2013). Long-term outdoor growth and lipid productivity of *Tetraselmis suecica, Dunaliella tertiolecta* and *Chlorella* sp. (Chlorophyta) in bag photobioreactors. *J. Appl. Phycol.* 25:167-176.

Moheimani, N. R. & Borowitzka, M. A. (2006). The long-term culture of the coccolothophorid *Pleurochrysis carterae* (Haptophyta) in outdoor raceway ponds. *J. Appl. Phycol.* 18:703-712.

Moheimani, N. R. & Borowitzka, M. A. (2007). Limits to productivity of the alga *Pleurochrysis carterae* (Haptophyta) grown in outdoor raceway ponds. *Biotechnol. & Bioengin.* 96:27-36.

Mohn, F. H. (1978). Improved technologies for the harvesting and processing of microalgae and their impact on production costs. *Archiv. Für Hydrobiologie Beihefte Ergebnisse der Limnologie* 1:228-253.

Moldowan, J. M. & Seifert, W. K. (1980). First discovery of botryococcane in petroleum. *J. S. C. Chem. Comm.* 912-914.

Molina Grima, E., Belarbi, E-H., Acien Fernandes F. G., Robles Medina, A. & Chisti, Y. (2003). Recovery of microalgal biomass and metabolites: process options and economics. *Biotechnol. Adv.* 20:491-515.

Monteiro, C. M., Fonseca, S. C., Castro, P. M. L. & Malcata, F. X. (2010). Toxicity of cadmium and zinc on two microalgae, *Scenedesmus obliquus* and *Desmodesmus pleiomorphus*, from Northern Portugal. *J. Appl. Phycol.* 23:97-103.

Montero, M. F., Aristizábal, M & Reina, G. G. (2011). Isolation of high-lipid content strains of the marine microalga *Tetraselmis suecica* for biodiesel production by flow cytometry and single-cell sorting. *J. Appl. Phycol.* 23:1053-1057.

Moore, J. K., Doney, S. C., Glover, D. M. & Fung, I. Y. (2002). Iron cycling and nutrient-limitation patterns in surface waters of the World Ocean. *Deep-Sea Res.* 49:463-507.

Moore, R. M., Webb, M., Tokarczyk, R & Wever, R. (1996). Bromoperoxidase and iodoperoxidase enzymes and production of halogenated methanes in marine diatom cultures. *J. Geophys. Res.* 101 (C9):20899-20908.

Mopper, K., Zhou, X., Kieber, R. J., Kieber, D. J. & Sikorski, R. J. (1991). Photochemical degradation of dissolved organic carbon and its impact on the ocean carbon cycle. *Nature* 353:60-62.

Morel, F. M. M. & Price, N. M. (2003). The biogeochemical cycles of trace metals in the oceans. *Science* 300:944-947.

Moreno-Garrido, I., Lubián, L. M. & Soares, A. M. V. M. (2000). Influence of cellular density on determination of EC_{50} in microalgal growth inhibition tests. *Ecotoxicol. Environ. Saf.* 47:112-116.

Morita, K., Ogata, M. & Hasegawa, T. (1999). *Chlorella* accelerates dioxin excretion in rats. *J. Nutrition* 129:1731-1736.

Morton, S. L., Moeller, P. D. R., Young, K. A. & Lanoune, B. (1998). Okadaic acid production from the marine dinoflagellate *Prorocentrum belizeanum* Faust isolated from the Belizean coral reef system. *Toxicon* 36:201-206.

Mukherjee, R., Borah, S. P. & Goswami, B. C. (2010). Biochemical characterization of carotenoids in two species of *Trentepohlia* (Trentepohliales, Chlorophyta). *J. Appl. Phycol.* 22:569-571.

Mulbry, W., Kebede-Westhead, E., Pizarro, C. & Sikora, L. J. (2005). Recycling of manure nutrients: use of algal biomass from dairy manure treatment as a slow release fertilizer. *Bioeesource Technol.* 96:451-458.

Mulbry, W., Kondrad, S. & Buyer, J. (2008). Treatment of dairy and swine effluents using freshwater algae: fatty acid content and composition of algal biomass at different loading rates. *J. Appl. Phycol.* 20:1079-1085.

Munné-Bosch, S. & Alegre, L. (2002). The function of tocopherols and tocotrienols in plants. *Crit. Rev. Plant Sci.* 21:31-57.

Muñoz, R. & Guieysse, B. (2006). Algal-bacterial processes for the treatment of hazardous contamination: a review. *Water Res.* 40:2799-2815.

Myklestad, S. M. (1995). Release of extracellular products by phytoplankton with special emphasis on polysaccharides. *Science of the Total Environment* 165:155-164.

Naar, J., Kubanek, J., Weidner, A. L., Flewelling, L. J., Bourdelais, A. J., Steidinger, K. A. & Baden, D. G. (2004). Brevetoxin depuration in shellfish via production of non-toxic metabolites: consequences for seafood safety and environmental fate of biotoxins. In: Steidinger, K. A. Landsberg, J. H. Tomas, C. R. & Vargo, G. A. (eds.) *Harmful Algae 2002.* Florida Fish and Wildlife Conservation Commission and Intergovernmental Oceanographic Commission UNESCO, St. Petersburg, pp. 488-490.

Naes, H. (1988). Geosmin production. *Methods in Enzymology* 167:605-608.

Naes, H., Aarnes, H., Utkilen, H. C., Nilsen, S. & Skulberg, O. M. (1985). Effects of photon fluence rate and specific growth rate on geosmin production of the cyanobacterium *Oscillatoria brevis* (Kütz) Gom. *Appl. Env. Microbiol.* 49:1538-1540.

Nagata, T. & Kirchman, D. L. (1992). Release of macromolecular organic complexes by heterotrophic marine flagellates. *Mar. Ecol. Prog. Ser.* 83:233-240.

Nakamura, K., Yamanaka, R., Tohi, K. & Hamada, H. (2000). Cyanobacterium-catalyzed asymmetric reduction of ketones. *Tetrahedron Lett.* 41:6799-6802.

Nakamura, N., Nakano, K., Sugiura, N. & Matsumura, M. (2003). A novel cyanobacteriolytic bacterium, *Bacillus cereus,* isolated from a eutrophic lake. *J. Biosci. Bioeng.* 95:179-184.

Nakajima, Y. & Ueda, R. (1997). Improvement of photosynthesis in dense microalgal suspensions by reduction of light harvesting pigment. *J. Appl. Phycol.* 9:503-510.

Nakajima, Y. & Ueda, R. (1999). Improvement of microalgal photosynthetic productivity by reducing the content of light harvesting pigment, *J. Appl. Phycol.* 11:195-201.

Nakazoe, J., Kimura, S., Yokoyama, M. & Iida, H. (1986). Effect of the supplementation of alga or lipids to the diets on the growth and body composition of nibbler, *Girella punctata* Grey. *Bull. Tokai Regional Fisheries Res. Lab.* 120:43-51.

Naumann, M. S., Richter, C., el-Zibdah, M. & Wild, C. (2009). Coral mucuc as an efficient trap for picoplanktonic cyanobacteria: implications for pelagic-benthic coupling in the reef ecosystem. *Mar. Ecol. Prog. Ser.* 385:65-76.

Negoro, M., Shioji, N., Miyamoto, K. & Miura, Y. (1991). Growth of microalgae in high CO_2 gas and effects of SO_x and NO_x *Appl. Biochem. Biotechnol.* 28-29:877-886.

Nell, J. A., Diemar, J. A. & Heasman, M. P. (1996). Food value of live yeasts and dry yeast-based diets fed to Sydney rock oyster, *Saccostrea commercialis* spat. *Aquaculture* 145:235-243.

Nell, J. A. & O'Connor, W. A. (1991). The evaluation of fresh algae and stored algal concentrates as a food source for Sydney rock oyster, *Saccostrea commercialis* (Iredale and Roughley), larvae. *Aquaculture* 99:277-284.

Nellemann, C., Corcoran, E. & Duarte, C. M. (Eds). (2009). *Blue Carbon: The Role of Healthy Oceans in Binding Carbon.* UNEP/Earthprint.

Neori, A. (2011). "Green water" microalgae: the leading sector in world aquaculture. *J. Appl. Phycol.* 23:143-149.

Newton, J. W., Tyler, D. D. & Slodki, M. E. (1979). Effect of ultraviolet-B 280-320 nm radiation on blue-green algae (cyanobacteria), possible biological indicators of stratospheric ozone depletion. *Appl. Environ. Microbiol.* 37:1134-1141.

Niki, T., Kunugi, M. & Otsuki, A. (2000). DMSP-lyase activity in five marine phytoplankton species: Its potential importance in DMS production. *Mar. Biol.* 136:759-764.

Nilsson, A., Petersson, F., Jönsson, H. & Laurell, T. (2004). Acoustic control of suspended particles in micro fluid chips. *Lab. Chip* 4:131-135.

Nizan, S., Dimentman, C. & Shilo, M. (1986). Acute toxic effects of the cyanobacterium *Microcystis aeruginosa* on *Daphnia magna. Limnol. Oceanogr.* 31:497-502.

Noda, K., Ohno, N., Tanaka, K., Kamiya, N., Okuda, M., Yadomae, T., Nomoto, K. & Shoyama, Y. (1996). A water-soluble antitumor glycoprotein from *Chlorella vulgaris. Planta Med.* 62:423-426.

North, F. K. (1985). *Petroleum Geology.* Chapter 7. The conversion of organic matter to petroleum. Allen and Unwin Inc., Boston, USA.

Nurdogan, Y. & Oswald, W. J. (1995). Enhanced nutrient removal in high-rate algal ponds. *Wat. Sci. Technol.* 31:33-43.

Oh, H-M., Yoo, C., Lee, J-Y. & Ahn, C. (2009). Physiological study of the increased lipid production of *Botryococcus braunii. Phycologia* 48, 280. 9th Inter. Phycol. Cong., Tokyo.

Ohtani, I., Moore, R. E. and others. (1992). Cylindrospermopsin: A potent hepatotoxin from the blue-green alga *Cylindrospermopsis raciborskii. J. Amer. Chem. Soc.* 114:7941-7942.

Oie, G., Makridis, P., Reitan, K. I. & Olsen, Y. (1997). Protein and carbon utilization of rotifers *Brachionus plicatilis*) in first feeding turbot larvae *Scophthalmua maximus* L. *Aquaculture* 153:103-122.

Okaichi, T. (1989). Red tide problems in the Seto Inland Sea, Japan. Pp 137- 142. In Okaichi, T., Anderson, D. M.& Nemoto, T (Eds.) *Red tides: Biology, environmental science and toxicology.* Elsevier, New York.

Olaizola, M. (2003). Microalgal removal of CO_2 from flue gases: changes in medium pH and flue gas composition do not appear to affect the photochemical yield of microalgal cultures. *Biotech. Bioproc. Eng.* 8:360-367.

Olguin, E. J. (2003). Phycoremediation key issues for cost-effective nutrient removal processes. *Biotechnol. Adv.* 22:81-91.

Olmedilla, B., Granado, F., Blanco, I & Vaquero, M. (2003). Lutein, but not &-tocopherol, supplementation improves visual function in patients with age –related cataracts: a 2y double-blind, placebo-controlled pilot study. *Nutrition* 19:21-24.

Oren, A. (2005). A hundred years of *Dunaliella* research: 1905-2005. *Saline Syst.* 1:2.

Oren, A. & Shilo, M. (1979). Anaerobic heterotrophic dark metabolism in the cyanobacterium *Oscillatoria limnetica*: sulphur respiration and lactate fermentation. *Arch. Microbiol.* 122:77-84.

Orr, J. C., Fabry, V. J., Aumont, O., Bopp, L and others. (2005). Anthropogenic ocean acidification over the twenty-first century and its impact on calcifying organisms. *Narure* 437:681-686.

Osborne, N. J. T., Webb, P. M. & Shaw, G. R. (2001). The toxins of *Lyngbya majuscula* and their human and ecological health effects. *Envir. International* 27:381-392.

Oshima, Y., Hasegawa, M., Yasumoto, T., Hallegraeff, G. & Blackburn, S. (1987). Dinoflagellate *Gymnodinium catenatum* as the source of paralytic shellfish toxins in Tasmanian shellfish. *Toxicon* 25:105-111.

Oswald, W.J. (1988). Microalgae and waste-water treatment In: Borowitzka, M. A. & Borowitzka, L. J. (eds.) *Microalgal Biotechnology,* Cambridge University Press, New York, USA. Pp305-328.

Palafox, N. A. & Buenconsejo-Lum, L. E. (2001). Ciguatera fish poisoning: review of clinical manifestations. *J. Toxicol Toxin Rev.* 20:141-160.

Panda, B., Sharma, L. & Mallick, N. (2005). Poly-ß-hydroxybutyrate accumulation in *Nostoc muscarum* and *Spirulina platensis* under phosphate limitation. *J. Plant Physiol.* 162;1376-1379.

Papandroulakis, N., Divanach, P. & Kentouri, M. (2002). Enhanced biological performance of intensive sea bream (*Spaus aurata*) larviculture in the presence of phytoplankton with long photophase. *Aquaculture,* 204:45-63.

Parameswari, E., Lakshmanan, A., & Thilagvathi, T. (2010). Phycoremediation of heavy metals in polluted water bodies. *Elec. J. Env. Agric. Food* 9:808-814.

Park, K. C., Whitney, C., McNichol, J. C. and others. (2012). Mixotrophic and photoautotrophic cultivationof 14 microalgae isolates from Saskatchewan, Canada: potential applications for wastewater remediation. *J. Appl. Phycol.* 24:339-348.

Partensky, F., Hess, W. R. & Vaulot, D. (1999). *Prochlorococcus,* a marine photosynthetic prokaryote of global significance. *Microbiol. Mol. Biol. Rev.* 63:106-127.

Passow, U. (2000). Formation of transparent exopolymer particles, TEP, from dissolved precursor material. *Mar. Ecol. Prog. Ser.* 192:1-11.

Peng, W., Wu, Q. & Tu, P. (2000). Effects of temperature and holding time on production of renewable fuels from pyrolysis of *Chlorella protothecoides. J. Appl. Phycol.* 12:147-152.

Peñueles, J. & Munné-Bosch, S. (2005). Isoprenoids: an evolutionary pool for photoprotection. *Rends in Plant Science,* 10:166-169.

Perez-Garcia, O., Bashan, Y. & Puente, M. E. (2011). Organic carbon supplementation of sterilized municipal wastewater is essential for heterotrophic growth and removing ammonium by the microalga *Chlorella vulgaris. J. Phycol.* 47:190-199.

Pesheva, I., Kodama, M., Dionisio-Sese, M. L. & Miyachi, S. (1994). Changes in photosynthetic characteristics induced by transferring air-grown cells of *Chlorococcum littorale* to high-CO_2 conditions. *Plant Cell Physiol.* 35:379-387.

Peters, J. W., Lanzilotta, W. N., Lemon, B. J. & Seefeldt, L. C. (1998) X-ray structure of the Fe-only hydrogenase (CpI) from *Clostridium pasteurianum* to 18 Å resolution .*Science* 282:1853-1858.

Petersson, F., Nilsson, A., Holm, C., Johnsson, H. & Laurell, T. (2005). Continuous separation of lipid particles from erythrocytes by means of laminar flow and acoustic standing wave forces. *Lab. Chip.* 5:20-22.

Petit, A-N., Eullaffroy, P., Debenest, T. & Gagné, F. (2010). Toxicity of PAMAM dendrimers to *Chlamydomonas reinhardtii*. *Aquatic Toxicology* 100:187-193.

Petukov, V. N., Fomchenkov, V. M., Chugunov, V. A. & Kholodenko, V. P. (2000). Plant biotests for soil and water contaminated with oil and oil products. *Appl. Biochem. & Microbiol.* 36:564-567.

Pfeiffer, T. J. & Ludwig, G. M. (2007). Small-scale system for the mass production of rotifers using algal paste. *North American J. Aquaculture* 69:239-243.

Phadwal, K. & Singh, P. K. (2003). Effect of nutrient depletion on β-carotene and glycerol accumulation in two strains of *Dunaliella* sp. *Bioresourc. Technol.* 90:55-58.

Picardo, M. C., de Medeiros, J. L., Monteiro, J. G. M., Chaloub, R. M., Giordano, M. & Araújo, O. deQ. F. (2012). A methodology for screening of microalgae as a decision making tool for energy and green chemical process applications. Clean Techn. Environ. Policy. DOI 10.1007/s10098-012-0508.

Picot, B., El Halouani, H., Casellas, C., Moersidik, S. & Bontoux, J. (1991). Nutrient removal by high rate pond system in a Mediterranean climate (France). *Wat. Sci. Technol.* 23:1535-1541.

Piiparinen, J. & Kuosa, H. (2011). Impact of UVA radiation on algae and bacteria in Baltic Sea ice. *Aquat. Mictob. Ecol.* 63:75-87.

Pimentel, D. (2008). A brief discussion on algae for oil production: Energy issues. In:Pimentel, D. (Ed.) *Biofuels, solar and wind as renewable energy systems.* pp499-500.Springer Science, USA.

Piorreck, M. & Pohl, P. (1984). Formation of biomass, total protein, chlorophylls, lipids and fatty acids in freshwater green and blue-green algae under different nitrogen regimes. *Phytochemistry* 23:207-216.

Pizarro, C., Mulbry, W., Blersch, D. & Kangas, P. (2006). An economic assessment of algal turf scrubber technology for treatmrnt of dairy manure effluent. *Ecol. Eng.* 26:321-327.

Pollio, A., Pinto, G., Della-Greca, M., Fiorentino, A. & Previtera, L. (1996). Biotransformations of progesterone by *Chlorella* spp. *Phytochemistry* 42:685-688.

Pollio, A., Pinto, G., Della Greca, M., Maio, A. D., Fiorentino, A. & Previtera, L. (1994). Progesterone bioconversion by microalgal cultures. *Phytochemistry* 37:1269-1272.

Poorvin, L., Rinta-Kanto, J. M., Hutchins, D. A. & Wilhelm, S. W. (2004). Viral release of iron and its bioavailability to marine plankton. *Limnol. Oceanogr.* 49:1734-1741.

Popovich, C. A., Damiani, C., Constenla, D. & Leonardi, P. I. (2012). Lipid quality of the diatoms *Skeletonema costatum* and *Navicula gregaria* from the South Atlantic Coast (Argentina): evaluation of its suitability as diodiesel feedstock. *J. Appl. Phycol.* 24:1-10.

Pörs, Y., Wüstenberg, A. & Ehwald, R. (2010). A batch culture method for microalgae and cyanobacteria with CO_2 supply throught polyethylene membranes. *J. Phycol.* 46:825-830.

Porter, D. (1990). Phylum Labyrinthulomycota. In: Margulis, L., Corliss, J. O., Melkonian, M. & Chapman, D. J. (Eds.). *Handbook of Protoctista.* Jones and Bartlett Publishers, Boston, pp. 388-398.

Posewitz, M. C., King, P. W., Smolinski, S. L., Zhang, L., Seibert, M. & Ghirardi, M. L. (2004). Discovery of two novel radical *S*-adenosylmethionine proteins required for the assembly of an active [Fe] hydrogenase. *J. Biol. Chem.* 279:25711-25720.

Posewitz, M. C., Smolinski, S. L., Kanakagiri, S., Melis, A., Seibert, M. & Ghirardi, M. L. (2004). Hydrogen photoproduction is attenuated by disruption of an isoamylase gene in *Chlamydomonas reinhardtii. The Plant Cell,* 16:2151-2163.

Poulsen, L. K., Moldrup, M., Berge, T. & Hansen, P. J. (2011). Feeding on copepod faecal pellets: a new trophic role of dinoflagellates as detritivors. *Mar. Ecol. Prog. Ser.* 441:65-78.

Prasanna, R., Singh, R. N. , Joshi, M., Madhan, K. Pal, R. K. & Nain, L. (2012). Monitoring the biofertilizing potential and establishment of inoculated cyanobacteria in soil using physiological and molecular markers. *J. Appl. Phycol.* 23:301-308.

Prasanna, R., Sood, A., Laiswal, P., Nayak, S., Gupta, V., Chaudhary, V., Joshi, M. & Natarajan, C. (2010). Rediscovering cyanobacteria as valuable sources of bioactive compounds (Review). *Appl. Biochem. Microbiol.* 46:119-134.

Pratt, R. (1944). Studies on *Chlorella vulgaris* IX. *Amer. J. Bot.* 31:418-420.

Price, G. D., Badger, M. R. & von Caemmerer, S. (2011). The prospect of using cyanobacterial bicarbonate transporters to improve leaf photosynthesis in C3 crop plants. *Plant Physiol.* 155:20-26.

Prince, R. C. & Kheshgi, H. S. (2005). The photobiological production of hydrogen: Potential efficiency and effectiveness as a renewable fuel. *Crit. Rev. Microbiol.* 31:1931.

Proctor, L.M. & Fuhrman, J. A. (1990). Viral mortality of marine bacteria and cyanobacyeria. *Nature* 343:60-62.

Proctor, V. W. (1957). Some controlling factors in the distribution of *Haematococcus pluvialis. Ecology* 38:457-462.

Przytocka-Jusiak, M., Blaszczyk, M., Kosińska, E. & Bisz-Konarzewska, A. (1984). Removal of nitrogen from industrial wastewaters with the use of algal rotating disks and denitrification packed bed reactor. *Wat. Res.* 18:1077-1082

Pugh, N., Ross, S. A., ElSohly, H. N., ElSohly, M. A. & Pasco, D. S. (2001). Isolation of three high molecular weight polysaccharide preparations with potent immunostimulotory activity from *Spirulina platensis, Aphanizomenon flos-aquae* and *Chlorella pyrenoidosa*. *Planta Med.* 67:737-742.

Pulz, O. (2001). Photobioreactors: production systems of phototrophic microorganisms. *Appl. Microbiol. Biotechnol.* 57:287-293.

Pulz, O. & Gross, W. (2004). Valuable products from biotechnology of microalgae. *Appl. Microbiol. Biotechnol.* 65:635-648.

Punshon, S. & Moore, R. M. (2008). Aerobic hydrogen production and dinitrogen fixation in the marine cyanobacterium *Trichodesmium erythraeum* IMS101.

Qiu, B. & Li, Y. (2006). Photosynthetic acclimation and photoprotective mechanism of *Haematococcus pluvialis* (Chlorophyceae) during the accumulation of secondary carotenoids at elevated irradiation. *Phycologia* 45:117-126.

Queiroz, M. I.,Lopez, E. J.,Zepka, L. Q., Bastos, R. G. & Goldbeck, R. (2007). The kinetics of the removal of nitrogen and organic matter from parboiled rice effeluent by cyanobacteria in a stirred batch reactor. *Bioresour. Technol.* 98:2163-2169.

Quilliam, M. A. (2003). Chemical methods for lipophilic shellfish toxins. In: Hallegraeff, G. M., Anderson, D. M. Cembella, A. D. (Eds.) *Manual on Harmful Marine Microalgae.* UNESCO, Paris.

Raja, R., Hemaiswarya, S., Kumar, N. A., Sridhar, S. & Rengasamy, R. (2008). A perspective on the biotechnological potential of microalgae. *Crit. Rev. Microbiol.* 34:77-88.

Ramos, A., Coesel, S., Marques, A., Rodrigues, M., Baumgartner, A., Noronha, J., Rauter, A., Brenig, B. & Varela, J. (2008). Isolation and characterization of a stress-inducible *Dunaliella salina Lcy-ß* gene encoding a functional lycopene ß-cyclase. *Appl. Microbiol. Biotechnol. 79:819-828.*

Ramos, M. J., Fernándes, C. M., Casas, A. Rodriguez, L. & Pérez, A. (2009). Influence of fatty acid composition of raw materials on biodiesel properties. *Bioresour. Technol.* 1:261-268.

Randt, C. & Senger, H. (1985). Participation of the two photosystems in light-dependent hydrogen evolution in *Scenedesmus obliquus. Photochemistry and Photobiology* 42:553-557.

Rapier, R. (2008). Renewable Diesel. In: Pimentel, D. (Ed.) *Biofuels, solar and wind as renewable energy systems.* pp. 153 *et seq*. Springer Science, USA.

Rasoul-Amini, S., Fotooh-Abadi, E. & Ghasemi, Y. (2012). Biotransformation of monoterpenes by immobilized microalgae. *J. Appl. Phycol.* 23:975-981.

Rasoul-Amini, S., Ghasemi, Y., Morowvat, M. H., Ghoshoon, M. B., Raee, M. J., Mosavi-Azam, S. B., Montazeri-Najafabady, N., Nouri, F., Parvizi, R., Negintaji, N. & Khoubani, S. (2010). Characterization of hydrocortisone bioconversion and 16S RNA gene in *Synechococcus nidulans* cultures. *Appl. Biochem. and Microbiol.* 46:191-197.

Ratledge, C. & Wynn, J. P. (2002). The biochemistry and molecular biology of lipid accumulation in oleaginous microorganisms. *Adv. Appl. Microbiol.* 51:1-43.

Reddy, C. S. K., Ghai, R. & Kalia, V. C. (2003). Polyhydroxyalkanoates: an overview. *Bioresour. Technol.* 87:137-146.

Reid, M. A., Tibby, J. C., Penny, D. & Gell, P. A. (1995). The use of diatoms to assess past and present water quality. *Aust. J. Ecol.* 20:57-64.

Reitan, K. I., Rainuzzo, J. R., Øie, G. & Olsen, Y. (1993). Nutritional effects of algal addition in first feeding of turbot *Scophthalmus maximus* L larvae. *Aquaculture* 118:257-275.

Rellán, S., Osswald, J., Saker, M., Gago-Martinez, A & Vasconcelos, V. (2009). First detection of anatoxin-a in human and animal dietary supplements containing cyanobacteria. *Food and Chemical Toxicology* 47:2189-2195.

Reshef, V., Mizrachi, E., Maretzki, T., Silberstein, C., Loya, S., Hizi, A. & Carmeli, S. (1997). New acylated sulfoglycolipids and digalactolipids and related known glycolipids from cyanobacteria with a potential to inhibit the reverse transcriptase of HIV-1. *J. Nat. Prod.* 60:1251-1260.

Revsbech, N.P., Jørgensen, B. B., Blackburn, T.H. & Cohen, Y. (1983). Microelectrode studies of the photosynthesis and O_2, H_2S and pH profiles of a microbial mat. *Limnol. Oceanogr.* 28:1062-1074.

Richlen, M. L. & Lobel, P. S. (2011). Effects of depth, habitat, and water motion on the abundance and distribution of ciguatera dinoflagellates at Johnston Atoll, Pacific Ocean. *Mar. Ecol. Prog. Ser.* 421:51-66.

Richmond, A. (1988). *Spirulina.* In: Borowitzka, A. & Borowitzka, L. (eds) *Microalgal Biotechnology.* Cambridge University Press pp. 83-121.

Richmond, A. (2004). Principles for attaining maximal microalgal productivity in photobioreactotrs. *Hydrobiologia* 512:33-37.

Richmond, A. & Grobbelaar, J. U. (1986). Factors affecting the output rate of *Spirulina platensis* with reference to mass cultivation. *Biomass* 10:253-264.

Richmond, A., Lichtenberg, E., Stahl, B. & Vonshak, A. (1990). Quantitative assessment of the major limitations on productivity of *Spirulina platensis* in open raceways. *J. Appl. Phycology* 2:195-206.

Riebesell, U. (2004). Effects of CO_2 enrichment on marine phytoplankton. *J. Oceanogr.* 60:719-729.

Riegger, L. & Robinson, D. (1997). Photoinduction of UV-absorbing compounds in Antarctic diatoms and *Phaeocystis Antarctica. Mar. Ecol. Prog. Ser.* 160:13-25.

Rise, M., Cohen, E., Vishkautsan, M., Cojocaru, M., Gottlieb, H. E. & Arad, S (M). (1994). Accumulation of secondary carotenoids in *Chlorella zofingiensis. J. Plant Physiol.* 144:287-292.

Roberts, B. S. (1979). Occurrence of *Gymnodinium breve* red tides along the west and east coasts of Florida during 1976 and 1977. Pp. 199-202. In Taylor, D. L. & Seliger, H. H. (eds.) *Toxic dinoflagellate blooms.* Elsevier/North Holland, New York.

Robineau, B., Gagné, J. A., Fortier, L. & Cembella, A. D. (1991). Potential impact of a toxic dinoflagellate (*Alexandrium excavatun*) bloom on survival of fish and crustacean larvae. *Mar. Biol.* (*Berl.*) 108:293-301.

Rocap, G., Larimer, F. W. and others. (2003). Genome divergence in two *Prochlorococcus* ecotypes reflects oceanic niche differentiation. *Nature* 424:1042-1047.

Rocchetta, I. & Küpper, H. (2009). Chromium- and copper-induced inhibition of photosynthesis in *Euglena gracilis* analysed on the single-cell level by fluorescence kinetic microscopy. *New Phytol.* 182:405-420.

Roessler, P. G. (1990). Environmental control of glycerolipids metabolism in microalgae: commercial implications and future research directions. *J. Phycol.* 26:393-399.

Roessler, P. G. & Lien, S. (1984). Purification of hydrogenase from *Chlamydomonas reinhardtii. Plant Physiol.* 75:705-709.

Roger, P. A. (1991). Reconsidering the utilization of blue-green algae in wetland rice cultivation. In: Dutta, S. K. & Sloger, C. (Eds.) *Biological Nitrogen Fixation Associated With Rice Production,* Oxford and IBH Publishing Co., New Delhi. pp 119-141.

Roger, P. A. & Kulasooriya, S. A. (1980). *Blue-green Algae and Rice,* Los Baňos, The Philippines: International Rice Research Institute.

Rohrlack, T., Christoffersen, K. & Friberg-Jensen, U. (2005). Frequency of inhibitors of daphnid trypsin in the widely distributed cyanobacterial genus *Planktothrix. Environ. Microbiol.*7:1667-1669.

Romay, C., Ledón, N. & Gonzalez, R. (2000). Effects of phycocyanin extract on prostaglandin E_2 levels in mouse ear inflammation tests. *Arzeinmittelforschung,* 50:1106-1109.

Round, F. E. (1991). Diatoms in river water-monitoring studies. *J. Appl. Phycol.* 3:129-145.

Round, F. E. (1993). A Review and Methods for the Use of Epilithic Diatoms for Detecting and Monitoring Changes in River Water Quality 1993. pp. 65. HMSO, London.

Ruff, T. A. (1989). Ciguatera in the Pacific: a link with military activities. *Lancet* 8631:201-204.

Runnegar, M. T., Kong, S. M., Zhong, Y. Z. & Lu, S. C. (1995). Inhibition or reduced glutathione synthesis by cyanobacterial alkaloid cylindrospermopsin in cultured rat hepatocytes. *Biochem.Pharmacol.* 49:219-225.

Saadoun, I. M. K., Schrader, K. K. & Blevins, W. T. (2001). Environmental and nutritional factors affecting geosmin synthesis by *Anabaena* sp. *Water Res.* 35:1209-1218.

Sabater, S., Gausch, H., Romani, A. & Muñoz, I. (2002). The effect of biological factors on the efficiency of river biofilms in improving water quality. *Hydrobiologia* 469:149-156.

Sabatini, S. E., Leonardi, P. I. & Rodriguez, M. C. (2011). Responses to sublethal copper exposure in two strains of *Chlamydomonas reinhardtii* (Volvocales,Chlorophyceae) in autotrophic and mixotrophic conditions. *Phycologia* 50:78 88. *

Sachs, J. P. & Repeta, D. J. (1999). Oligotrophy and nitrogen fixation during eastern Mediterranean sapropel events. *Science* 286:248.

Saker, M.L., Moreira, C., Martinus, J., Neilan, B. & Vasconcelos, V. M. (2009). DNA profiling of complex bacterial populations: toxic cyanobacterial blooms. *Appl. Microbial. Biotechnol.* DOI 10.1007/s00253-009-2180-8.

Saker, M. L., Welker, M., & Vasconcelos, V.M. (2007). Multiplex PCR for the detection of toxigenic cyanobacteria in dietary supplements produced for human consumption. *Appl. Microbiol. & Biotechnol.* 73:1136-1142.

Sandmann, G, (2001) Biosynthesis of carotenoids – *Arch Biochim. Biophys.* 385:4-12.

Sano, T., Takagi, H., Morrison, L. F. Metcalf, J. S. Codd, G. A. & Kaya, K. (2005). Leucine aminopeptidase M inhibitors, cyanostatin A and B, isolated from cyanobacterial water blooms in Scotland. *Phytochemistry* 66:543-548.

Santos-Rosa, F., Galván, F. & Vega, J. M. (1989). Biological viability of *Chlamydomonas reinhardtii* cells entrapped in alginate beads for ammonium photoproduction. *J. Biotechnol.* 9:209-219.

Santoya, S., Jaime, L., Plaza, M and others. (2012). Antiviral compounds obtained from microalgae commonly used as carotenoid sources. *J. Appl. Phycol.* 24:731-741.

Sato, N., Tsuzuki, M., Matsuda, Y., Ehara, T., Osafune, T. & Kawaguchi, A. (1995). Isolation and characterization of mutants affected in lipid metabolism of *Chlamydomonas reinhardtii. Eur. J. Biochem.* 230:987-993.

Satoh, G., Urashima, A., Kurosaki, D., Kato, S., Shinoharra, T., Inamori, Y. & Yukawa, Y. (2009). Introduction of the new microalgal strain having a capacity for efficient CO_2 removal. 9th Int. Phycol. Congr. Tokyo, Japan, *Phycologia* 48, Abstract No. 340, p. 119.

Sattler, S. E., Gilliland, L. U., Magallanes-Lundback, M., Pollard, M. & DellaPenna, D. (2004). Vitamin E is essential for seed longevity and for preventing lipid peroxidation during germination. *Plant Cell,* 16:1419-1432.

Schantz, E. J., Lynch, J.M., Vayvada, G., Matsumoto, K. & Rapoport, H. (1966). The purification and characterization of the poison produced by *Gonyaulax catenella* in axenic culture. *Biochemistry* 5:1191-1195.

Schantz, E. J., Mold, J. D., Stanger, D. W., Shavel, J., Riel, F. J., Bowden, J. P., Lynch, J. M., Wyler, R. S., Reigel, B. & Sommer, H. (1957). Paralytic shellfish poisoning, VI. A procedure for the isolation and purification of the poison from toxic clams and mussel tissues. *J. Am. Chem. Soc.* 78:5230-5233.

Schenk, P. M., Thomas-Hall, S. R., Stephens, E., Marx, U. C., Mussgnug, J. H., Posten, C., Kruss, O. & Hankamer, B. (2008). Second generation biofuels: high-efficiency microalgae for biodiesel production. *Bioenerg. Res.* 1:20-43.

Schirmer, A., Rude, M. A., Li, X.Z., Popova, E. & del Cardayre, S. B. (2010). Microbial biosynthesis of alkanes. *Science* 329:559-562.

Schlegel, I., Doan, N. T., de Chazal, N. & Smith, G. D. (1999). Antibiotic activity of new cyanobacterial isolates from Australia and Asia against green algae and cyanobacteria. *J. Appl. Phycol.* 10:471-479.

Schmidt, L. E. & Hansen, P. J. (2001). Allelopathy in the prymnesiophyte *Chrysochromulina polylepis*: effect of cell concentration, growth phase and pH. *Mar. Ecol. Prog. Ser.* 216:67-81.

Schnackenberg, J., Schulz, R. & Senger, H. (1993). Characterization and purification of a hydrogenase from the eukaryotic green alga *Scenedesmus*. *FEBS Lett.* 327:21-24.

Schrader, K. K. (2007). Preliminary pond evaluation of a natural-based algicide for managing musty off-flavours in channel catfish aquaculture. *North American J. Aquaculture* 69:360-366.

Schrader, K. K., Rimando, A. M., Tucker, C. S., Glinski, J., Cutler, S. J. & Cutler, H. G. (2004). Evaluation of the natural product SeaKleen for controllong the musty-odor-producing cyanobacterium *Oscillatoria perornata* in catfish ponds. *North American J. Aquaculture* 66:20-28.

Schrader, K. K., Tucker, C. S., Hanson, T.R., Gerard, P.D., Kingsbury, S. K. & Rimando, A. M. (2005). Management of musty off-flavour in channel catfish from commercial ponds with weekly application of copper sulphate. *North American J. Aquaculture* 67:138-147.

Schrauzer, G. N. (2000). Anticarcinogenic effects of selenium. *Cell. Mol. Life Sci.* 57:1864-1873.

Schwarz, M. H., Craig, S. R., Delbos, B. C. & McLean, E. (2008). Efficacy of concentrated algal paste during greenwater phase of cobia larviculture. *J. Appl. Aquaculture* 20:285-294.

Seckbach, J., Baker, A. F. & Shugarman, P. M. (1970). Algae thrive under pure CO_2 *Nature* 227:744-745.

Seikai, T., Tanangonan, J. & Tanaka, M. (1986). Temperature influence on larval growth and metamorphosis of the Japanese flounder *Paralichthys olivaceus* in the laboratory. *Bull. Japanese Soc. Sci. Fisheries* 52:977-982.

Semple, K. T., Cain, R. B. & Schmidt, S. (1999). Biodegradation of aromatic compounds by microalgae. *FEMS Microbiol. Lett.* 170:291-300.

Senger, H. & Bishop, N.I. (1979). Observations on the photohydrogen producing activity during the synchronous cell cycle of *Scenedesmus obliquus*. *Planta* 145:53-62.

Sepúlveda, J., Wendler, J. E., Summons, R. E. & Hinrichs, K-U. (2009). Rapid resurgence of marine productivity after the Cretaceous-Paleogene mass extinction. *Science* 326:129-131.

Setlik, I., Veladimir, S. & Malek, I. (1970). Dual purpose open circulation units for large scale culture of algae in temperate zones. I. Basic design considerations and scheme of pilot plant. *Algol. Stud. (Trebon)* 1:11.

Shashirekha, S., Uma, L. & Subramanian, G. (1997). Phenol degradation by the marine cyanobacterium *Phormidium valderianum* BDU 30501. *J. Ind. Microbiol. Biotechnol.* 19:130-133.

Sheehan, J., Dunahay, T., Benemann, J. & Roessler, P. (1998). A look back at the U.S. Department of Energy's Aquatic Species program: biodiesel from algae. *National Renewable Energy Laboratory Report* NREL/TP-580-24190.

Shelly, K., Heraud, P. & Beardall, J. (2002). Nitrogen limitation in *Dunaliella tertiolecta* (Chlorophyceae) leads to increased susceptibility to damage by ultraviolet-B radiation but also increased repair capacity. *J. Phycol.* 38:713-720.

Shen, X., Lam, P. K., Shaw, G. R & Wickramasinghe, W. (2002). Genotoxicity investigation of a cyanobacterial toxin, cylindrospermopsin. *Toxicon* 40:499-501.

Shi, J., Podola, B. & Melkonian, M. (2007). Removal of nitrogen and phosphorus from wastewater using microalgae immobilized on twin layers: an experimental study. *J. Appl. Phycol.* 19:417-423.

Shifrin, N. S. & Chisholm, S. W. (1981). Phytoplankton lipids:interspecific differences and effects of nitrate, silicate and light-dark cycles. *J. Phycol.* 17:374-384.

Shimamatsu, H. (2004). Mass production of *Spirulina,* an edible microalga. *Hydrobiologia,* 512:39-44.

Shimizu, Y. & Li, B. (2006). Microalgae as a source of bioactive molecules: special problems and methodology. In: *Frontiers in marine biotechnology* (Ed. P. Proksch & W. E. G. Muller) pp.145-174. Horizon Bioscience, Wymonham, UK.

Siegenthaler, U & Sarmiento, J. L. (1993). Atmospheric carbon dioxide and the ocean. *Nature* 365:119-259.

Sidari, L., Nichetto, P., Cpk, S., Sosa, S., Tubaro, A., Honsell, G. & Della Loggia, R. (1998). Phytoplankton selection by mussels, and diarrhetic shellfish poisoning. *Marine Biology* 131:103-111.

Sieracki, M., Poulton, N. & Crosbie, N. (2005). Automated isolation techniques for microalgae. In: *Algal culturing techniques* (Ed. R. A. Anderson), pp.101-116. Elsevier Academic Press, New York.

Simmons, T. L. & Gerwick, W. H. (2008). Anticancer drugs of marine origin. In: Walsh, P., Smith, S, Fleming, L.Solo-Gabriele, H., Gerwick, W. H. (eds.) *Oceans and human health: risks and remedies from the seas.* Academic Press, New York, pp 431-452.

Simó, R., Archer, S. D., Pedrós-Alió, C., Gilpin, L. & Stelfox-Widdicombe, C. E. (2002). Coupled dynamics of dimethylsulfoniopropionate and dimethylsulfide cycling and the microbial food web in surface waters of the North Atlantic. *Limnol. Oceanogr.* 47:53-61.

Simó, R., Pedrós-Alió, C., Malin, G. & Grimalt, J. O. (2000). Biological turnover of DMS and DMSO in contrasting open-sea waters. *Mar. Ecol. Prog. Ser.* 203:1-11.

Simon, C. M. (1978). The culture of the diatom *Chaetoceros gracilis* and its use as a food for penaeid protozoeal larvae. *Aquaculture* 14:105-113.

Sims, P. A., Mann, D. G. & Medlin, L. K. (2006). Evolution of the diatoms: insights from fossils, biological and molecular data. *Phycologia* 45:361-402.

Sinha, R. P. & Häder, D. P. (2000). Effects of UV-B radiation on cyanobacteria. Recent Research Dev. *Photochem. Photobiol.* 4:239-246.

Sinigalliano, C. D., Winshell, J., Guerrero, M. A., Scorzetti, G., Fell, J. W., Eaton, R. W., Brand, L. & Rien, K. S. (2009). Viable cell sorting of dinoflagellates by multiparametric flow cytometry. *Phycologia* 48:249-257.

Sirin, S., Trobajo, R., Ibanez, C. & Salvadó, J. (2011). Harvesting the microalga *Phaeodactylum tricornutum* with polyaluminium chloride, aluminium sulphate, chitosan and alkalinity-induced flocculation. *J. Appl. Phycol.* DOI 10.1007/s10811-011-9736-6.

Sládečková, A. (1994). The role of periphyton in wastewater technology. *Ver.Int. Verein. Limnol.* 25:1929-1932.

Smayda, T. J (1991). Global epidemic of noxious phytoplankton blooms and food chain consequences in large ecosystems. In Sherman, K., Alexander, L. M. & Gold, B. D. (Eds.) *Food chains, yields, models and management of large marine ecosystems.* Westview Press, Boulder, USA.

Smayda, T. J. (1997). Harmful algal blooms: their ecophysiology and general relevance to phytoplankton blooms in the sea. *Limnol. Oceanogr.* 42:1137-1153.

Smetacek, V. (1999). Diatoms and the ocean carbon cycle. *Protist* 150:25-32.

Smith, R. C., Prezelin, B. B., Baker, K. S., Bidigare, R. R., Boucher, N. P., Coley, T. Karentz, D., MacIntyre, S., Matlick, H. A., Menzies, D., Ondrusek, M., Wan, Z. & Waters, K. J. (1992). Ozone depletion: Ultraviolet radiation and phytoplankton biology in Antarctic waters. *Science* 255:952-959.

Søderberg, L. M. & Hansen, P. J. (2007). Growth limitation due to high pH and low inorganic carbon concentrations in temperate species of the dinoflagellate genus *Ceratium. Mar. Ecol. Prog. Ser.* 351:103-112.

Soeder, C. J. (1980). Massive cultivation of microalgae: Results and prospects. *Hydrobiologia* 72:197-209.

Sohrin, R., Isaji, M., Obara, Y., Agostini, S., Suzuki, Y., Hiroe, Y., Ichikawa, T. & Hidaka, K. (2011). Distribution of *Synechococcus* in the dark ocean. *Aquat. Microb. Ecol.* 64:1-14.

Solomon, S., Mills, M., Heidt, L. E., Pollock, W. H. & Tuck, A. F. (1992). On the evaluation of ozone depletion potentials. *J. Geophys. Res.* 97:825-842.

Solovchenko, A. E., Khozin-Goldberg, I., Didi-Cohen, S., Cohen, Z. & Mwerzlyak, M. N. (2008). Effects of light intensity and nitrogen starvation on growth, total fatty acids and arachidonic acid in the green microalga *Parietochloris incisa. J. Appl. Phycol.* 20:245-251.

Somogyi, B., Felföldi, T., Solymosi, K., Makk, J., Homonnay, Z. G., Horváth, G., Turcsi, E., Böddi, B., Márialigeti, K. & Vörös, L. (2011). *Chloroparva pannonica gen. et sp. nov.* (Trebouxiophyceae, Chlorophyta) – a new picoplanktonic green alga from a turbid, shallow soda pan. *Phycologia* 50:1-10.

Spencer, A. M., LecKie, G. G. & Chew, K. J. (1996). North Sea hydrocarbon plays and their resources. In: Glennie, K. & Hurst, A. (eds.) *Northwest Europes's Hydrocarbon Industry.* Geological Society, London. Pp. 25-41.

Spilling, K., Seppälä, J. & Tamminen, T. (2010). Inducing autoflocculation in the diatom *Phaeodactylum tricornutum* through CO_2 regulation. *J. Appl. Phycol.* DOI:10.1007/s10811-010-9616-5.

Spoehr, H. A. & Milner, H. W. (1949). The chemical composition of *Chlorella;* effects of environmental conditions. *Plant Physiol.* 24:120-149.

Spolaore, P., Joannis-Cassen, C., Duran, E. & Isambert, A. (2006). Commercial applications of microalgae. *J. Biosci. Bioeng.* 101:87-96.

Spruit, C. J. P. (1958). Simultaneous photoproduction of hydrogen and oxygen by *Chlorella. Meded. Landbouwhogeschool, Wageningen.*58:1-17.

Stal, L. J. & Moezelaar, R. (1997). Fermentation in cyanobacteria. *FEMS Microbiology Reviews* 21:179-211.

Steinbrenner, J. & Linden, H. (2001). Regulation of two carotenoid biosynthesis genes coding for phytoene synthase and carotenoid hydroxylase during stress-induced astaxanthin formation in the green alga *Haematococcus pluvialis. Plant Physiol.* 125:810-817.

Stoermer, E. F., Wolin, J. A., Schelske, C. L. & Conley, D. J. (1985). An assessment of ecological changes during recent history of Lake Ontario based on sileous algal microfossils preserved in sediments. *J. Phycol.* 21:257-276.

Strmečki, S., Plavšić, M., Steigenberger, S. & Passow, U. (2010). Characterization of phytoplankton exudates and carbohydrates in relation to their complexation of copper, cadmium and iron. *Mar. Ecol. Prog. Ser.* 408:33-46.

Stuart, T. S. & Kaltwasser, D. (1970). Photoproduction of hydrogen by photosystem I of *Scenedesmus. Planta* 91:302-313.

Stuart, T. S. & Gaffron, H. (1972). The mechanism of hydrogen photoproduction by several algae. II. The contribution of photosystem II. *Planta* 106:101-112.

Su, C-S., Chien, L-J., Gomez, J., Lin, Y-S., Yu, Y-K., Liou, J-S. & Syu, R-J. (2011). Factors affecting lipid accumulation by *Nannochloropsis oculata* in a two-stage cultivation process. *J. Appl. Phycol.* 23:903-908.

Suárez, E. R., Kralovec, J. A., Noseda, M. D.,Ewart, H. S., Barrow, C. J., Lumsden, M. D. & Grindley, T. B. (2005). Isolation, charcterization and structural determination of a unique type of arabinogalactan from an immunostimulatory extract of *Chlorella pyrenoidosa. Carbohydr. Res.* 340:1489-1498.

Suárez, E. R., Syvitski, R., Kralovec, J. A., Noseda, M. D., Barrow, C J., Ewart, H. S., Lumsden, M. D. & Grindley, T. B. (2006). Immunostimulatory polysaccharides from *Chlorella pyrenoidosa.* A new galactofuranan. Measurement of molecular weight and molecular weight dispersion by DOSY NMR. *Biomacromolecules* 7:2368-2376.

Sukenik, A. (1999). Production of EPA by *Nannochloropsis.* In: Cohen, Z. (ed.) *Chemicals from microalgae.* Taylor & Francis, London, pp41-56.

Sukenik, A. & Shelef, G. (1984). Algal autoflocculation – verification and proposed mechanism. *Biotechnol. Bioeng.* 26:142-147.

Sunda, W. G. (1995). The influence of nonliving organic matter on the availability and cycling of plant nutrients in seawater. In: Zepp, R. G. & Sonntag, C. (eds.).*Role of nonliving organic matter in the Earth's carbon cycle.* John Wiley & Sons, Chichester, pp.191-207.

Sunda, W. G., Kieber, D. J., Kiene, R. P. & Huntsman, S. (2002). An antioxidant function for DMSP and DMS in marine algae. *Nature* 418:317-320.

Surai, P. F. (2006). Selenium in nutrition and health. Nottingham University Press, Nottingham. Pp. 974.

Suto, I. (2007). The Oligocene and Miocene record of the diatom resting spore genus *Liradiscus* Greville in the Norwegian Sea. *Micropaleontology* 53:145-159.

Suttle, C. A. (2000). Cyanophages and their role in the ecology of cyanobacteria. In: *The Ecology of Cyanobacteria: Their Diversity in Time and Space* Whitton, B. A. & Potts, M. (eds.), pp.563-589. Kluwer Academic Publishers, Boston.

Suttle, C. A. (2007). Marine viruses – major players in the global ecosystem. *Nature Rev. Microbiol.* 5:801-812.

Suttle, C. A. & Chan, A. M. (1994). Dynamics and distribution of cyanophages and their effects on marine *Synechococcus* spp. *Appl. Environ. Mircobiol.* 60:3167-3174.

Suzuki, E., Shiraiwa, Y. & Miyachi, S. (1994). The cellular and molecular aspects of carbonic anhydrase in photosynthetic microorganisms. *Progress in Phycological Research* 10:1-54.

Tadonléké, R. D. (2010). Evidenace of warming effects on phytoplankton productivity rates and their dependence on eutrophic status. *Limnol. Oceanogr.* 55:973-982.

Tahedl, H. & Häder, D. P. (1999). Fast examination of water quality using the automatic biotest ECOTOX based on movement behaviour of a freshwater flagellate. *Water Res.* 33:426-432.

Takagi, M., Karseno, & Yoshida, T. (2006). Effects of salt concentration on intracellular accumulation of lipids and triacylglyceride in marine microalgae *Dunaliella* cells. *J. Biosci. Bioeng.* 101:223-226.

Takahashi, Y. (1985). Morphological and behavioural changes with growth in reared larvae and juveniles of a flounder, *Paralichthys olivaceus. Aquaculture* 33:43-52.

Talukdar, J., Kalita, M. C. & Goswami, B. C. (2009). Oleaginous microalgae of Northeastern India: Potential alternatives for sustainable development of biofuels. *Phycologia* 48:127.

Tamagnini, P., Leitão, E., Oliveira, P., Ferreira, D., Pinto, F., Harris, D. J., Heidorn, T. & Lindblad, P. (2007). Cyanobacterial hydrogenases: diversity, regulation and applications. *FEMS Microbial Rev.* 31:692-720.

Tamelander, T., Aubert, A. B. & Riser, C. W. (2012). Export stoichiometry and contribution of copepod faecal pellets to vertical flux of particulate organic carbon, nitrogen and phosphorus. *Mar. Ecol. Prog. Ser.* 459:17-28.

Tan, L. T. (2007). Bioactive natural products from marine cyanobacteria for drug discovery. *Phytochemistry* 68:954-979.

Tanoi, T., Kawachi, M & Watanabe, M. M. (2010). Effects of carbon source on growth and morphology of *Botryococcus braunii. J. Appl. Phycology.* DOI 10.1007/s10811-010-9528-4.

Tappan, H. & Loeblich, A. R. Jr. (1973). Evolution of the oceanic plankton. *Earth-Science Reviews* 9:207-240.

Tatters, A. O., Muhlstein, H. I. & Tomas, C. R. (2010). The haemolytic activity of *Karenia selliformis* and two clones of *Karenia brevis* throughout a growth cycle. *J. Appl. Phycol.* 22:435-442.

Tenaud, M., Ohmori, M. & Miyachi, S. (1989). Inorganic carbon and acetate assimilation in *Botryococcus braunii* (Chlorophyta). *J. Phycol.* 25:662-667.

Thauer, R. K., Jungermann, K. & Decker, K. (1977). Energy conservation in chemotrophic anaerobic bacteria. *Bacteriol. Rev.* 41:100-180.

Thompson, P.A., Guo, M. & Harrison, P. J. (1992). Effects of variation in temperature. I. On the biochemical composition of eight species of marine phytoplankton. *J. Phycol.* 28:481-488.

Tjahjono, A. E., Hayama, Y., Kakizono, T., Terada, Y., Nishio, N. & Nagai, S. (1994). Hyper-accumulation of astaxanthin in a green alga *Haematococcus pluvialis* at elevated temperatures. *Biotechnol. Letts.* 16:133-138.

Tocher, D. R., Leaver, M. J. & Hodgson, P. A. (1998). Recent advances in the biochemistry and molecular biology of fatty acyl desaturases. *Prog. Lipid Res.* 37:73-117.

Tolleter, D. and others. (2011). Control of hydrogen photoproduction by the proton gradient generated by cyclic electron flow in *Chlamydomonas reinhardtii The Plant Cell* 23:2619-2630.

Tredici, M. R. (2004). Mass production of microalgae: photobioreactors. In: Richmond, A. (ed). *Handbook of microalgal culture.* Blackwell, Oxford. pp. 178-214.

Tremblay, P., Naumann, M. S., Sikorski, S., Grover, R. & Ferrier-Pages, C. (2012). Experimental assessment of organic carbon fluxes in the scleractinian coral *Stylophora pistillata* during a thermal and photo stress event. *Mar. Ecol. Prog. Ser.* 453:63-77.

Tsoglin, L. N. & Gabel', B. V. (2000). The technology of production of biomass labelled with stable isotopes. Abstract from 4th European Workshop on Biotechnology of Microalgae. Berghholz Rehbrucke, Germany.

Tsygankov, A. A. (2001). Laboratory Scale Photobioreactors. *Appl. Biochem. Microbiol.* 37:333-341.

Tsygankov, A. A., Serebryakova, L. T., Rao, K. K. & Hall, D. O. (1998). Acetylene reduction and hydrogen photoproduction by wild-type and mutant strains of *Anabaena* at different CO_2 and O_2 concentrations. *FEMS Microbiol. Letts.* 167:13-17.

Turpin, D. H., Layxell, D. B. & Elrifi, I. R. (1985). Modeling the carbon economy of *Anabaena flos-aquae. Plant Physiol.* 78:746-752.

Tyagi, R., Kumar, A., Tyagi, M. B.,Jha, P. N., Kumar, H. D., Sinha, R. P. & Häder, D. P. (2003). Protective role of certain chemicals against UV-B-induced damage in the nitrogen-fixing cyanobacterium, *Nostoc muscarum. J. Basic Micrbiol.* 43:137-147.

Tyagi, R., Srinivas, G., Vyas, D., Kumar, A. & Kumar, H. D. (1992). Differential effect of ultraviolet-B radiation on certain metabolic processes in a chromatically adapting *Nostoc. Photochem. Photobiol.* 55:401-407.

Uribe, J. C. & Ruiz, M. (2001). *Gymnodinium* brown tide in the Magellanic fjords, southern Chile. *Rev. Biol. Mar. Oceanogr.* 36:155-164.

Vandamme, D., Foubert, I., Meesschaert, B. &.Muylaert, K. (2010). Flocculation of microalgae using cationic starch. *J. Appl. Phycol.* 22:525-530.

Van den Hoek, C., Mann, D. G. & Jahns, H. M. (1995). *Algae, an introduction to phycology.* Cambridge University Press, Cambridge, England.

Van Gemerden, H. (1993). Microbial mats: a joint venture. *Mar. Geol.* 113:3-25.

Van Krevelen, D. W. (1981). *Coal – typology, chemistry, physics, constitution.* Elsevier Scientific Publishing Co. Amsterdam, New York.

Van Landingham, S. L. (1964). Chrysophyta cysts from the Yakima Basalt (Miocene) in south-central Washington. *J. Paleontology* 31:729-739.

Van Landingham, S. L. (1991). Precision dating by means of traditional biostratigraphic methods for the middle Miocene diatomaceous interbeds within the middle Yakima (Wanapum) Basalt of south-central Washington (U.S.A.). *Nova Hedwigia* 53:349-368.

Van Landingham, S. L. (2010). Use of diatoms in determining age and paleoenvironment of the Valsequillo (Hueyatlaco) early man site, Puebla, Mexico, with corroboration by Chrysophyta cysts for a maximum Yarmouthian (430,000-500,000 yr BP) age of the artefacts. *Nova Hedwigia,* 136:127-138.

Vanucci, S., Dell'Anno, A., Pusceddu, A., Fabiano, M., Lampitt, R. S. & Danovaro, R. (2001). Microbial assemblages associated with sinking particles in the Porcupine Abyssal Plain (NE Atlantic Ocean). *Prog. Oceanogr.* 50:105-121.

Varfolomeev, S. D. & Wasserman, L. A. (2011). Microalgae as source of biofuel, food, fodder, and medicines. *Appl. Biochem. & Microbiol.* 47:789-807.

Vaulot, D., Eikrem, W., Viprey, M. & Moreau, H. (2008). The diversity of small eukaryotic phytoplankton (< or = 3 microm.) in marine ecosystems. *FEMS Microbiol. Rev.* 32:795-820.

Villar, R., Laguna, M. R., Calleja, J. M. & Cadavid, I. (1992a). Effects of *Skeletonema costatum* extracts on the central nervous system. *Planta Med.* 58:398-404.

Villar, R., Laguna, M. R., Calleja, J. M. & Cadavid, I. (1992b). Effects of *Phaeodactylum tricornutum* and *Dunaliella tertiolecta* extracts on the central nervous system. *Planta Med.* 58:405-409.

Villar, R., Laguna, M. R., Cadavid, I. & Calleja, J. M. (1994). Effects of aqueous extracts of six marine microalgae on smooth muscle contraction in rat duodenum and vas deferens. *Planta Med.* 60:521-526.

Vincent Inc. (2007). Valcent Products. Initial data from the Vertgro Field Test Bed Plant reports average production of 276 tons of algae biomass on a per acre/year basis. From http://money.cnn.com/newsfeeds/articles/marketwire/033918.htm

Vincent, W. F. & Roy, S. (1993). Solar UV-B effects on aquatic primary production: damage, repair and recovery. *Environ. Rev.* 1:1-12.

Visviki, L. & Rachlin, J. W. (1991). Ultrastructural changes in *Dunaliella minuta* following acute and chronic exposure to copper and cadmium. *Arch. Environ. Contam. Toxicol.* 23:420-425.

Volbeda, A., Charon, M. H., Piras, C., Hatchikian, E. C., Frey, M. & Fontecilla-Camps, J. C. (1995). Crystal structure of the nickel-iron hydrogenase from *Desulfovibrio gigas*. *Nature* 373:580-587.

Volk, R-B., Venzke, K. & Blaschek, W. (2007). Structural investigation of a polysaccharide released by the cyanobacterium *Nostoc insulare*. *J. Appl. Phycol.* 19:255-262.

Volk, T. (1989). Sensitivity of climate and atmospheric CO_2 to deep ocean- and shallow-ocean carbonate burials. *Nature* 337:637-640.

Vorst, P., Baard, R. L., Mur, L.R., Korthals, H. J. & van den Ende, H. (1994). Effect of growth arrest on carotene accumulation and photosynthesis in *Dunaliella*. *Microbiology,* 140:1411-1417.

Voushak, A. & Richmond, A. (1988). Mass production of bluegreen alga *Spirulina*: an overview. *Biomass* 15:233-247.

Vymazal, J. (1988). The use of periphyton communities for nutrient removal from polluted streams. *Hydrobiologia* 166:225-237.

Wada, H. & Murata, N. (1989). *Synechocystis* PCC6803 mutants defective in desaturation of fatty acids. *Plant Cell Physiol.* 30:971-978.

Wang, B., Karka, A. et al. (2003). Astaxanthin accumulation in *Marmatococcus pluvialis* (Chlorophyceae) as an active photoprotective process under high irradiance. *J. Phycol.* 39:1116-1124.

Wang, B., Li, Y. Q., Wu, N. & Lan, C. Q. (2008). CO_2 bio-mitigation using microalgae. *Appl. Microbiol. Biotechnol.* 79:707-718.

Wang, L-C., Wu, J-T., Lee, T-Q. & Chen, S-H. (2009). Use of diatoms in lake sediments to infer habitat changes and fluctuation in precipitation over late Holocene in Eastern Taiwan. *Phycologia* 48 Abstract 52. 9th Internat. Phycol. Congr. Tokyo.

Wängberg, S-A. (1995). Effects of arsenate and copper on the algal communities in polluted lakes in the northern parts of Sweden assayed by PICT (Pollution-Induced Community Tolerance). *Hydrobiologia* 306:109-124.

Ward, O. P. & Singh, A. (2005). Omega-3/6 fatty acids: alternative sources of productions. *Proc. Biochem.* 40:3267-3652.

Warthmann, R., Pfennig, N. & Cypionka, H. (1993). The quantum requirement for H_2 production by anoxygenic phototrophic bacteria. *Appl. Microbiol. Biotechnol.* 39:358-362.

Watanabe, A., Ito, R. & Konishi, C. (1951). Effect of nitrogen-fixing blue-green algae on the growth of rice plants. *Nature* 168:748-749.

Watanabe, J., Natsume, T. & Kobayashi, M. (2007). Comparison of the antivascular and cytotoxic activities of TZT-1027 (soblidotin) with those of other anticancer agents. *Anticancer Drugs* 18:905-911.

Watanabe, T., Kazumi, A. & Akiko, H. (1988). Numerical water quality monitoring of organic pollution using diatom assemblages. In: *Proceedings of the Ninth International Diatom Symposium: Bristol, August 24-30, 1986.* (ed. F. E. Round) pp.123-141. Biopress Ltd.., Bristol and Koeltz Scientific Books, Koenigstein.

Watanabe, T., Kitajima, C. & Fujita, S. (1983). Nutritional values of live organisms used in Japan for mass propagation of fish: a review. *Aquaculture* 34:115-143.

Watson, A. S. (1979). Aquaculture and algae culture. Processes and production. Noyes Data Corporation, NJ.

Weckesser, J. & Drews, G. (1979). Lipopolysaccharides of photosynthetic prokaryotes. *Ann. Rev. Microbiol.* 33:215-239.

Weetall, H. H. (1985). Studies on the nutritional requirements of the oil-producing alga *Botryococcus braunii. Appl. Biochem. Biotech.* 11:377-391.

Welch, I. M., Barrett, P. R. F., Gibson, M. T. & Ridge, I. (1990). Barley straw as an inhibitor of algal growth. I Studies in the Chesterfield Canal. *J. Appl. Phycol.* 2:231-239.

White, A. W. (1984). Paralytic shellfish toxins and finfish. Pp. 171-180. In Ragelis, E. P. (Ed.) *Seafood toxins.* ACS Symposium Series No. 262, Washington, D.C. American Chemical Society.

White, A. W. (1977). Dinoflagellate toxins as probable cause of an Atlantic herring (*Clupea harengus harengud*) kill, and pteropods as apparent vector. *J. Fish. Res. Board Can.* 34:2421-2424.

White, A. W., Fukuhara, O. & Anraku, M. (1989). Mortality of fish larvae from eating toxic dinoflagellates or zooplankton containing dinoflagellate toxins. Pp. 395-398. In: Okaichi, T , Anderson, D. M. & Nemoto, T. (Eds.). *Red tides: Biology, environmental science and toxicology.* Elsevier, New York.

Whitton, B. A. & Potts, M. (2000). *The Ecology of Cyanobacteria: their diversity in time and space.*Kluwer Academic Publishers, Dordrecht, The Netherlands.

Whyte, J. N. C. (1987). Biochemical composition and energy content of six species of phytoplankton used in mariculture of bivalves. *Aquaculture,* 60:231-241.

Wijffels, R. H. & Barbosa, M. J. (2010). An outlook on microalgal biofuels. *Science* 329:796-799.

Wilde, E. W. & Benemann, J. R. (1993). Bioremoval of heavy metals by the use of microalgae. *Biotechnol. Adv.* 11:781-812.

Wilkinson, S. C., Goulding, K. H. & Robinsin, P. K. (1990). Mercury removal by immobilized algae in batch culture systems. *J. Appl. Phycol.* 2:223-230.

Willatts, P. & Forsyth, J. S. (2000). The role of long chain polyunsaturated fatty acids in infant cognitive development. *Prostaglandins Leukot. Essent. Fatty Acids* 63:95-100.

Wilson, A., Boulay, C., Wilde, A., Kerfeld, C. A. & Kirilovsky, D. (2007). Light-induced energy dissipation in iron-starved cyanobacteria: Roles of OCP and IsiA proteins. *Plant Cell* 19:656-672.

Wilson, S. E., Cumming, B. F. & Smol, J. P. (1994). Diatom-salinity relationships in 111 lakes from the Interior Plateau of British Columbia, Canada: The development of diatom-based models for palaeosalinity reconstructions. *J. Palaeolimnol.* 12:197-221.

Wingfield, G. I., Greaves, M. P., Bebb, J. M. & Seager, M. (1985). Microbial immobilisation of phosphorus as a potential means of reducing phosphorus pollution of water. *Bull. Environ. Contam. Toxicol.* 34:587-596.

Winter, A., Jordan, R.W. & Roth, P. H. (1994). Biogeography of living coccolithophores in ocean waters. In: Winter, A. & Siesser, W. G. (eds.). *Coccolithophores.* Cambridge University Press, Cambridge, pp. 161-177.

Wolf, F. R., Nonomura, A. M. & Bassham, J. A. (1985). Growth and branched hydrocarbon production in a strain of *Botryococcus braunii* (Chlorophyta). *J. Phycol.* 21:388-396.

Wolfe, G. V. (2000). The chemical defense ecology of marine unicellular plankton: constraints, mechanisms and impacts. *Biol. Bull.* 198:225-244.

Wu, G. F., Wu, Q. Y. & Shen, Z. Y. (2001). Accumulation of poly-ß-hydroxybutyrate in cyanobacterium *Synechocystis* sp. PCC6803. *Bioresour. Technol.*76:85-90.

Wünschiers, R. (2006). Nitrogen availability for nitrogen fixing cyanobacteria upon growth on dinitrogen. *African J. Biotechnol.* 5:1969-1972.

Wünschiers, R., Senger, H. & Schulz, R. (1998). Photohydrogen production – hydrogenases in the green alga *Scenedesmus obliquus*. In: Garab, G. (Ed.) *Photosynthesis: Mechanisms and Effects, Vol. III.* Kluwer Academic Publishers, Netherlands. pp. 1951-1954.

Wurster, M., Mundt, S., Hammer, E., Schauer, F. & Lindequist, U. (2003). Extracellular degradation of phenol by cyanobacterium *Synechococcus* PCC 7002. *J. Appl. Phycol.* 15:171-176.

Wykoff, D. D., Davies, J. P., Melis, A. & Grossman, A. R. (1998). The regulation of photosynthetic electron transport during nutrient deprivation in *Chlamydomonas reinhardtii*. *Plant Physiol.* 117:129-139.

Xie, P. (1999). Gut contents of silver carp, *Hypophthalmichthys molitrix* and the disruption of a centric diatom, *Cyclotella,* on passage through the oesophagus and intestine. *Aquaculture* 180:295-305.

Xiong, W., Li, X., Xiang, J. & Wu, Q. (2008). High-density fermentation of microalga *Chlorella protothecoides* in bioreactor for microbio-diesel production. *Appl. Microbiol. Biotechnol.* 78:29-36.

Xiong, W., Liu, L., Wu, C., Yang, C. & Wu, Q. (2010). ^{13}C-Tracer and gas chromatography-mass spectrometry analyses reveal metabolic flux distribution in the oleaginous microalga *Chlorella protothecoides*. *Pl. Physiol.* 154:1001-1011.

Xiong, W., Li, X., Xiang, J & Wu, K. Y. (2008). High-density fermentation of microalga *Chlorella protothecoides* in bioreactor for microdiesel production. *Appl. Microbiol. Biotechnol.* 78:29-36.

Xue, L., Zhang, Y., Zhang, T., An, L. & Wang, X. (2005). Effects of enhanced Ultraviolet-B radiation on algae and cyanobacteria. *Crit. Revs. In Microbiol.,* 31:79-89.

Yamaguchi, K. (1997). Recent advances in microalgal bio-science in Japan, with special reference to utilization of biomass and metabolites: a review. *J. Appl. Phycol.* 8:487-502.

Yang, J-R. & Dickman, M. (1993). Diatoms as indicators of lake trophic status in central Ontario, Canada. *Diatom Res.* 8:179-193.

Yan, R., Zhu, D., Zhang, Z., Zeng, Q. & Chu, J. (2012). Carbon metabolism and energy conversion of *Synechococcus* sp. PCC 7942 under mixotrophic conditions: comparison with photoautotrophic condition. *J. Appl. Phycol.*24:659-668.

Yan, X., Chen, D., Xu, J. & Zhou, C. (2011). Profiles of photosynthetic glycerolipids in three strains of *Skeletonema* determined by UPLC-Q-TOF-MS. *J. Appl. Phycol.* 23:271-282.

Yasumoto, T., Murata, M., Oshima, Y., Sano, M., Matsumoto, G. K. & Clardy, J. (1985). Diarrhetic shellfish toxins. *Tetrahedron* 41:1019-1025.

Yen, T. F. & Chilingarian, G. V. (1976). *Oil Shale* pp.292. Elsevier, Amsterdam.

Yoon, K., Han, D, Li, Y., Sommerfield, M & Hu, Q. (2012). Phospholipid:Diacylglycerol Acyltransferase is a multifunctional enzyme involved in lipid turnover and degaradation while synthesising triacylglyserol in the unicellular green microalga *Chlamydomonas reinhardtii. The Plant Cell,* 24:3708-3724.

Young, J. R., Bown, P. R. & Burnett, J. A. (1994). Palaeontological perspectives. In: Green, J. C. & Leadbeater, B. S. C. (Eds.) *The Haptophyte Algae.* Systematics Association Special Volume No. 51. Clarendon Press, Oxford. pp. 379–392.

Yousoff, F. M., Matias, H. B., Khalid, Z. A. & Phang, S-M. (2001). Culture of microalgae using interstitial water extracted from shrimp pond bottom sediments. *Aquaculture* 201:263-270.

Yu, E. T., Zendejas, F. J., Lane, P. D., Gaucher, S., Simmons, B. A. & Lane, T. W. (2009). Triacylglycerol accumulation and profiling in the model diatoms *Thalassiosira pseudonana* and *Phaeodactylum tricornutum* (Baccilariophyceaea) during strarvation. *J. Appl. Phycol.* 21:669-681.

Yu, H. (2013). Effect of mixed carbon substrate on exoplolysaccharide production of cyanobacterium *Nostoc flagelliforme* in mixotrophic cultures. *J. Appl. Phycol.* 24:669-673.

Zakharova, Y. R. & Parfenova, U. V. (2007). A method for cultivation of microorganisms oxidising iron and manganese in bottom sediments of Lake Baikal. *Biol. Bull.* 34:236-241.

Zamalloa, C., Vrieze, J. D., Boon, N. & Verstraete, W. (2012). Anaerobic digestibility of marine microalgae *Phaeodactylum tricornutum* in a lab-scale anaerobic membrane bioreactor. *Appl. Microbiol. Biotechnol.* 93:859-869.

Zamani, N., Noshadi, M., Amin, S., Niazi, A. & Ghasemi, Y. (2012). Effect of alginate structure and microalgal immobilization method on orthophosphate removal from wastewater. *J. Appl. Phycol.* 24:649-656.

Zaslavskaia, L. A., Lippmeier, J. C., Shih, C., Ehrhardt, D., Grossman, A. R. & Apt, K. (2001). Trophic conversion of an obligate photoautotrophic organism through metabolic engineering. *Science* 292:2073-2075.

Zeng, D., Wu, J. & Kennedy, J. F. (2008). Application of a chitosan flocculant to water treatment. *Carbohyd. Polym.* 71:135-139.

Zehr, J. P., Waterbury, J. B., Turner, P. J., Montoya, J. P., Omoregie, E., Steward, G. F., Hansen, A. & Karl. D. M. (2001). Unicellular cyanobacteria fix N_2 in the subtropical North Pacific Ocean. *Nature* 412:635-638.

Zhang, H., Hu, C., Jia, X., Xu, Y., Wu, C., Chen, L. & Wang, F. (2012). Characteristics of γ-hexachlorocyclohexane biodegradation by a nitrogen-fixing cyanobacterium, *Anabaena azotica*. *J. Appl. Phycol.* 24:221-225.

Zhang, L. & Melis, A. (2002). Probing green algal hydrogen production. *Phil. Trans. R. Soc. Lond.* B. 357:1499-1509.

Zhang, S., Xu, C. & Santschi, P. H. (2008). Chemival composition and ^{234Th} (IV) binding of extracellular polymeric substances (EPS) produced by the marine diatom *Amphora* sp. *Marine Chemistry* 112:81-92.

Zhang, T., Li, L., Song, L. & Chen, W. (2009). Effects of temperature and light on the growth and geosmin production of *Lyngbya kuetzingii* (Cyanobacteria). *J. Appl. Phycol.* 21:279-285.

Zhang, H. Z., Pan, W-b. & Ma, C. (2010). Isolation and identification of the extracellular algae-lysing components from algae-lysing bacterial strain L7. *China Environ. Sci 30 (Suppl)*:19-23 (in Chinese).

Zhao, S., Pan, W-b. & Ma, C. (2012). Stimulation and inhibition effects of algae-lytic products from *Bacillus cereus* strain L7 on *Anabaena flos-aquae*. *J. Appl. Phycol.* 23:1015-1021.

Zielinski, U. & Gersonde, R. (1997). Diatom distribution in Southern Oceanic sediments (Atlantic sector): implications for palaeoenvironmental reconstructions. *Palaeogeogr. Palaeoclimat. Palaeoecol.* 129:213-250.

Zondervan, I. (2007). The effects of light, macronutrients, trace metals and CO_2 on the production of calcium carbonate and organic carbon in coccolithophores – a review. *Deep Sea Res.II* 54:521-537.

Index

#

A

B

C

D

E

F

G

H

I

J

K

L

M

N

O

P

Q

R

S

T